J.P.Morgan

OVERLOADED

Also available in the Bloomsbury Sigma series:

Sex on Earth by Jules Howard
Spirals in Time by Helen Scales
A is for Arsenic by Kathryn Harkup
Suspicious Minds by Rob Brotherton
Herding Hemingway's Cats by Kat Arney
The Tyrannosaur Chronicles by David Hone
Soccermatics by David Sumpter
Big Data by Timandra Harkness
Goldilocks and the Water Bears by Louisa Preston
Science and the City by Laurie Winkless
Built on Bones by Brenna Hassett
The Planet Factory by Elizabeth Tasker
Catching Stardust by Natalie Starkey
Seeds of Science by Mark Lynas
Nodding Off by Alice Gregory
The Science of Sin by Jack Lewis
The Edge of Memory by Patrick Nunn
Turned On by Kate Devlin
Borrowed Time by Sue Armstrong
The Vinyl Frontier by Jonathan Scott
Clearing the Air by Tim Smedley
Superheavy by Kit Chapman
The Contact Paradox by Keith Cooper
Life Changing by Helen Pilcher
Sway by Pragya Agarwal
Bad News by Rob Brotherton
Kindred by Rebecca Wragg Sykes
Mirror Thinking by Fiona Murden
Our Only Home by His Holiness The Dalai Lama
First Light by Emma Chapman
Ouch! by Margee Kerr & Linda Rodriguez McRobbie

OVERLOADED

How every aspect of your life is influenced by your brain chemicals

Ginny Smith

BLOOMSBURY SIGMA
LONDON • OXFORD • NEW YORK • NEW DELHI • SYDNEY

BLOOMSBURY SIGMA
Bloomsbury Publishing Plc
50 Bedford Square, London, WC1B 3DP, UK
29 Earlsfort Terrace, Dublin 2, Ireland

BLOOMSBURY, BLOOMSBURY SIGMA and the Bloomsbury Sigma logo are trademarks of Bloomsbury Publishing Plc

First published in the United Kingdom in 2021

A catalogue record for this book is available from the British Library

Library of Congress Cataloguing-in-Publication data has been applied for

ISBN: HB: 978-1-4729-6934-7; TPB: 978-1-4729-6935-4;
eBook: 978-1-4729-6936-1

2 4 6 8 10 9 7 5 3 1

Typeset by Deanta Global Publishing Services, Chennai, India
Printed and bound in Great Britain by CPI Group (UK) Ltd, Croydon CR0 4YY

Illustration by Marc Dando

Bloomsbury Sigma, Book Sixty-three

Contents

CHAPTER ONE

The Chemical Brain

[illegible]

CHAPTER ONE

The Chemical Brain

This is a book about some of the most fundamental questions in neuroscience; perhaps in science as a whole. How does our brain produce our everyday experiences, and drive our behaviour? How can a kilo and a half of jelly allow us to learn a foreign language, decide what to have for lunch, or even fall in love? While the cells that make up our brains are important for this, it is the chemicals that bathe them, and allow them to communicate, that paint the complex details which colour every aspect of our daily lives. But how exactly can these tiny molecules cause the full spectrum of human experience, with all its richness, its highs and lows, its joys and sorrows? And how does the brain ensure it isn't overloaded by this melee of molecules?

In this book, we will uncover cutting-edge research and meet leading scientists aiming to better understand the complex and intricate workings of your brain, and the molecules that control it. We will explore drugs, both medicinal and recreational, which alter the levels of these molecules, and investigate how a better understanding of our brain's workings might help us improve treatments for common conditions, without overloading the delicate balance of our chemical brain. We will even touch on ideas about free will, consciousness, and how our brain enables us to control our instincts.

Along the way we will dig into the history of neuroscience, uncovering stories of scientists' curiosity and persistence in the face of huge challenges. There are also tales of accidental discoveries which have revolutionised our understanding of the brain. I find these stories of how science is done as fascinating as the science itself, as they help us get a deeper understanding of the way knowledge develops over the years as well as giving us a glimpse into the characters behind the chemicals.

My hope is that this book arms you against the over-simplification rife in the media. As we will see, the idea that 'serotonin is the happiness chemical' or 'dopamine is addictive' misses so much of the important nuance as to render these statements pointless. Instead, here, we will explore that complexity and celebrate it, keeping things comprehensible and cutting through the scientific jargon to examine the underlying concepts. While this book may not be able to provide all the answers to these fundamental questions (in many cases we just don't know them yet), I hope that it sparks your curiosity, and encourages you to want to find out more about your incredible brain.

More questions than answers

I have always been curious, wanting to understand why and how the things around me work. As a child I was lucky to have parents who encouraged this curiosity, and did their best to answer my questions. In one infamous story, my mum remembers us walking into a public toilet and me, aged around three, turning to her and asking: 'Mummy, why are sounds louder in here?'

As she began explaining to me how sound is absorbed by soft surfaces, but bounces off hard ones, of which there are far more in a public toilet, another mum came out of the cubicle and gave her an admiring (or perhaps astonished) look, saying: 'I'm glad my daughter doesn't ask me questions like that!'

I always loved science, soaking up knowledge and understanding, but it wasn't until university that I discovered my fascination with brain science. As so often seems to happen, the decisions that change the course of our lives aren't those huge, momentous ones we agonise over, but the seemingly small, inconsequential ones. And this was definitely the case for me. It was just a couple of lectures, given as part of a course called 'Evolution and Behaviour', which I picked on a bit of a whim,* that opened my eyes to the wonders of the human brain. Here was a real challenge. An incredibly complex system, full of mysteries and unknowns, that needed breaking down, and understanding at the most fundamental level. And not only that, but it was something that affected us all, every day. I was hooked, and decided to change the focus of my degree from chemistry to psychology.

Over the summer before my second year, worried that my lack of biology A-level would cause problems for me, I refreshed my limited knowledge of the nervous system. I had learnt a little about the structure of neurons, or nerve cells, at GCSE. I knew they were specialised

* And, if I'm honest, because I thought it would give me the excuse to watch lots of David Attenborough documentaries as 'revision'.

cells that send messages around the brain and body, but I needed to learn more. So I began reading.

Neurons, I discovered, come in a range of types. In the body, there are sensory neurons, which carry information from your senses to the central nervous system, which is made up of your brain and spinal cord. Motor neurons carry information in the opposite direction, allowing your brain to control your movements. Then there are tiny interneurons, which connect the two and allow complex circuits to form. In the brain, things are more complicated, and we can't categorise neurons quite so neatly, as they come in so many different shapes and have so many different functions. But there are some similarities between neurons in the brain and the body.

Just like most other cells, neurons have cell bodies. These contain a nucleus (which has a range of roles, including storing the DNA) and mitochondria (which produce energy). It is in this cell body that new proteins are made, allowing the cell to function, and repair itself when needed. But it was the differences between neurons and 'typical' animal cells that I found most fascinating: the dendrites, tree-like branches, reaching out from the neuron to allow it to connect with many other neurons, and the long 'axons' that allow it to send messages quickly and easily.

Jennifer Aniston and Frankenstein's frogs

Our understanding of how messages pass along neurons began in 1780, with a scientist called Luigi Galvani. Trained in medicine and surgery, after graduating Galvani became a lecturer at the University of Bologna. Alongside his teaching, he carried out research, and developed an interest in the ways in which electricity could affect the body. He discovered that when electricity was passed through the leg of a dead frog, it began to twitch, as if it was coming back to life.* Galvani was amazed, and named his discovery 'animal electricity'. He believed he had found the animal's 'life force', and, to some extent at least, he was right. But he was wrong to think that 'animal electricity' was a specialised form of electricity. Alessandro Volta, a professor of experimental physics in the University of Pavia, Italy, realised this, and publicly criticised Galvani's theory. He proved that animal electricity was the same form of electricity that flowed through the fluid in his 'voltaic pile', which went on to become the batteries we still use today. In both the battery and the body, electricity is carried by a flow of charged particles, or ions.

We now know that when activated, electricity travels from one end of a neuron to the other,† a bit like a crowd doing a Mexican wave at a football match. This is known as a 'spike' or 'action potential', and is often referred to as the neuron 'firing'. Action potentials are all or nothing, more like a digital watch where it can be either 3.46 or 3.47 than an analogue one where the hands can point to anywhere in between. You can't have a small or large action potential; either the neuron reaches its threshold and fires, or it doesn't. So rather than the strength of firing, it is the rate of the spikes that

---

* This is thought to have inspired Mary Shelley's novel *Frankenstein*, written in 1818.

† Interestingly, in most neurons, information can only flow in one direction: the first neuron releases the chemicals and the second has the receptors to receive them.

provides information about, for example, the intensity of a sensation.

As I delved deeper, I realised how complex these seemingly simple structures are. The axon, for example, is coated with a fatty substance called myelin. This acts a bit like the insulation you find around electrical cables, but also helps messages to be sent more quickly along axons. This myelin is what is sometimes referred to as 'white matter' in the brain and is made by glial cells.* Although there are ten times as many glial cells as neurons in the brain, they are often ignored in favour of their more famous cousins. This is, at least partly, because they used to be thought of as fairly inert, just there to support the neurons. We are now starting to realise that glial cells take a much more active role in keeping the brain healthy and functioning properly.

The more I learnt about the complexities and intricacies of our brains, the more hooked I became. In my final year of university, I began to realise my interest lay in the boundary between psychology and neuroscience, an area known as cognitive psychology or behavioural neuroscience. This was the area, it seemed to me, that could answer the most important question: why do we behave the way we do? But the deeper I got into it, the more I realised how much we don't know. For many years, after the invention of MRI scanners, which allowed scientists to peer inside a living human brain, there was a focus on finding out which areas of the brain are involved in which processes. Scientists found areas 'for' numbers, faces and laughing when being tickled.

* Glial cells are any cells in the brain that aren't neurons.

They even found a single neuron which responded to pictures of Jennifer Aniston. But more recently, we have started to realise that this is an over-simplification, caused by the resolution of brain scanners. Rather than looking at areas 'for' each process, we now search for networks: combinations of neurons which can be distributed widely across the brain.

In a human brain, each neuron can make thousands of connections, building up the dense web of cells which makes up your brain's grey matter. It is these connections between cells that I find particularly fascinating, because this is what gives our brains their amazing flexibility. We are born with the vast majority of our neurons already in place. There is some debate over whether certain regions of the brain can make new neurons, but this process doesn't seem to be widespread. But the neurons we already have *can* make new connections, and the *strength* of connections can change. Sometimes this happens via neurons growing new dendrites, and new physical connections being made. But this is a slow process, and our brains need a faster way to change the signals that are sent around them. This is where brain chemicals come in. But to understand how these work, we first need to look at how they were discovered – and that means stepping back in time again, this time to the mid-1800s.

The argument of the century

In the middle of the nineteenth century, what brains were made of was still mysterious. Human senses are only so good, and the microscopes of the day weren't hugely powerful. What was clear was that the brain was a network, consisting of densely packed, intertwining fibres. But what these fibres were, and how they were responsible for the myriad of duties our brain carries out every second we are alive was far from clear. Of course, this didn't stop scientists putting forward theories and carrying out experiments to try to determine the truth. And one of these scientists was Camillo Golgi.

Golgi was born in Italy in 1843 and grew up to follow his father into medicine. His interest in experimental medicine, however, took him back to a research position and he began to study under Giulio Bizzozero. Amongst other things, Bizzozero was an expert in the nervous system, and studied its structure under a microscope.

At this time, the cell theory of physiology was still relatively young. While the idea that the body consisted of discrete elements, or cells, had been widely accepted, it hadn't been extended to the brain. Instead, German anatomist Joseph von Gerlach put forward reticular theory, suggesting the fibres of the nervous system formed a continuous network, through which fluid, carrying information, could flow. It was this theory Golgi supported, and this shaped his findings throughout his life.

When money issues forced him to take a job in a hospital once more, Golgi refused to give up his research, setting up a crude laboratory in an old hospital kitchen. In order to see the nervous system under a microscope, scientists of the time created stains, which aimed to dye these structures and make them more visible. But Golgi felt they weren't good enough to make the complex structures of the brain clear. Frustrated, he set about experimenting, and in 1873, he made a breakthrough. By treating brain specimens first with potassium dichromate and then silver nitrate, he found that just a few of the densely packed brain cells turned black, making their structure easily visible under a microscope. He named his technique *la reazione nera* (the black reaction), but today it is better known as the Golgi stain, and it is still in widespread use. Golgi published many papers using his stain to support the reticular hypothesis, arguing it allowed us to see the complex network through which information flowed. But there was a controversy brewing, one which would continue to rage for the rest of Golgi's life. And to understand that, we need to travel across the Mediterranean Sea, to Spain.

In the 1860s, while Golgi was still completing his studies, a Spanish anatomy teacher and his wife were desperately worried about their young son. Kicked out of one school after another for poor behaviour, the young Santiago Ramón y Cajal had even been in trouble with the law for building a cannon that destroyed a neighbour's gate. After apprenticeships to a shoemaker and a barber did little to curb his anti-authoritarian attitude, his father spent a summer taking him to graveyards to find human remains they could sketch, hoping to spark an interest in medicine. A keen artist growing up, Cajal took to this immediately, and in 1873 he graduated in medicine, before becoming an army medic. This was short-lived, however, as bouts of malaria and tuberculosis contracted in Cuba brought him back to Spain where, after recovering from his illnesses, he turned his attention to research.

Cajal's early work focused on inflammation, cholera and the study of the skin, but in 1887, in Barcelona, he heard about Golgi's new staining method, which sparked an interest in the brain that would continue throughout his life. Cajal was not just a scientist but also an artist, and his drawings of neurons are as beautiful as they are detailed. And it was while sitting at the microscope, making these incredible drawings, that Cajal realised Golgi had got something wrong. The brain wasn't one continuous web, but was made up of discrete cells. This provided support for what was known as the neuron theory. Despite the pair sharing the Nobel Prize in 1906, Golgi never accepted Cajal's evidence for neuron theory, and continued to support the reticular theory until his death in 1926.

Of course, modern imaging techniques have proved without a doubt that Cajal and other proponents of the neuron theory were correct. Neurons don't quite connect, leaving a small gap between the end of one and the start of the next. This gap is called a synapse, and is vitally important in the way our brain works.

Mind the gap!

In some synapses, the electrical current can pass straight from the first (or presynaptic) neuron to the second via something called a gap junction. But much more common in the adult brain are chemical synapses, and it is these we will be focusing on in this book.

When a neuron is activated, as we have seen, an electrical signal flows along it from one end to the other. When this action potential reaches the end, it causes chemicals called neurotransmitters to be released. These travel across the synapse until they reach the next neuron. At the very edge of this neuron are special structures called receptors, which sit half in and half out of the cell. When the chemicals bind to these receptors, they cause changes in the second cell. What changes they cause depends both on the chemical released and on the type of receptor, but they fall into two main categories. The first is fast and direct, and the second slower, operating through a cascade of messengers within the neuron.

In the first category, when the neurotransmitters bind to the receptors, they directly change the flow of ions inside the neuron. This can make the cell more 'excitable', and, if enough neurotransmitters bind, can cause the neuron to fire. If a neurotransmitter has this effect, we call it excitatory. Through a similar process, neurotransmitters can have the opposite effect, changing the flow of ions to make it harder to activate a neuron. These are inhibitory neurotransmitters.

These processes are relatively fast and short-lasting, but the second type of receptor allows chemicals to have slower and longer-lasting effects, triggering a cascade of changes within the cell. These can work to make the cells easier or harder to activate, change the way neurotransmitters are released, or alter the receptors found in the neuron. There can be multiple types of these receptors for each neurotransmitter, meaning the same chemical can have different effects depending on which receptors it binds to.

The most common neurotransmitter in the human brain is called glutamate. In most cases, glutamate has an excitatory effect on the second neuron, making it easier for the signal to be sent and activating the second cell. If enough glutamate is released, the signal is transmitted from the first cell to the second, and can continue on its journey. As we'll see, this chemical is vital for all sorts of processes, from learning to pain.

Found widely throughout the brain, Gamma Aminobutyric Acid, known as GABA, has the opposite effect to glutamate. When it binds to receptors, it makes it harder for the neuron to send an action potential. This makes GABA an inhibitory neurotransmitter. Neurons that use GABA, then, have a calming effect on the brain, reducing the activity of other neurons. This means they are important for sleep, and for counteracting anxiety.

Most of the brain chemicals we come across in this book, however, won't be this simple. They may have different effects in different parts of the brain, based on which types of receptors are found there. There can even be multiple types of receptor found on the same neuron, so the same chemical may have different effects depending on how much of the chemical is released, or how long for. This might sound confusing, but as we will discover, each process alone is relatively simple, and it is by combining them that we reach the greater levels of complexity that make our brain such an amazing machine.

Then there is the question of reach. In the simple examples of neurotransmission we have discussed here, GABA or glutamate is released into a synapse and affects one or maybe a couple of neurons that are on the other side of it. But sometimes brain chemicals are released more widely and affect a whole group of neurons, making it easier or harder for them to pass messages between them. This extra level of complexity is what allows our brains to change so quickly, and gives us such amazing flexibility in our behaviour.

This brings us back to that fundamental question, the one that hooked me all those years ago. How do our brains make us behave the way we do? It seems to me that the answer lies not in the wiring of our brains, but in the chemicals that bathe them. Because while, as we will see, the connections between neurons can and do change,* this process is slow. This means it can't be responsible for the millisecond-by-millisecond changes we all experience: the split-second decisions, fluctuations in emotion and the temptations we encounter. Instead, these are all controlled by our brain chemistry.

And, as it is our brain that makes us who we are, that means that *we* are controlled by this turbulent sea of neurotransmitters. But how does this work? To find out, I have spent the last 18 months reading books and papers and talking to experts around the world. And while I have found some answers, I have also discovered a myriad of questions. Again and again I have run up against the boundaries of scientific knowledge, and rediscovered just how much there is still to learn about our incredible, complex brains. Neuroscience moves faster, I would argue, than any other science, so while I have attempted to present the current scientific consensus in each topic, there will be people who disagree. And, by the time you are reading this, new studies will have been published that might turn our understanding on its head again. That is the way science works, so it is best to treat this book as a snapshot of our understanding at this moment in time, not as facts, carved in stone for evermore.

Each chapter in this book covers a different aspect of our lives, and how they are affected by our brain

---

* And this process itself is controlled by chemicals, as we will see in Chapter 2.

chemistry. These are huge topics, and each could have been (and has been) the focus of a book in its own right, so I have had to pick and choose which stories to cover. This means there will, unavoidably, be gaps – topics and areas that could have been included in each chapter, but which I haven't been able to go into. I have provided a section with further reading suggestions for each chapter at the end of the book, should you want to delve deeper into any one area.

I have attempted to provide the chapters in an order that makes sense, with each building on those that have come before. The intricate relationship between brain areas, networks and chemicals means that there is more in common than you might expect between topics that may seem, on the surface, completely different. You can, of course, dip into any chapter, and read them in whatever order you choose, and each should make sense alone. But I encourage you, if you can, to start at the beginning, because, as you will see, the brain is highly interconnected.

Now, without further ado, I invite you to join me on a journey of discovery, and find out more about the chemical soup that makes you you…

CHAPTER TWO

Thanks for the Memories

Do you remember what you were doing on your 18th birthday? Mine was a Friday and after college my boyfriend drove me, along with a couple of friends, to Reading town centre. It was a chilly January evening as we walked to the Purple Turtle, a characterful (if a touch grimy) cocktail bar a little way off the high street. There, proudly wearing my '18 today' badge, I enjoyed my new-found ability to buy alcohol legally by working my way through their extensive cocktail menu. My memories of this evening (well, the early parts of it at least!) are fairly clear. I even remember queuing for the toilet, where a group of girls asked me if it really was my birthday.*

But if you picked another Friday from that year and asked me what I did, I wouldn't have a clue. On a superficial level, it seems obvious that momentous occasions are stored differently, so they can be recalled much more easily than other, more humdrum days. But how is this difference coded in our brains? What is actually going on in my neurons and synapses when I recall that evening that brings the sticky floors and graffiti-covered walls to mind so readily?

---

* Which did raise the question, do people really wear '18 today' badges on days other than their 18th birthday? And if so, why?

To understand that, we have to look at how we learn and remember, a journey that will introduce us to the ethical quagmire that is cognitive enhancing drugs and make us question the reliability of our own memories. But first, let's start with the basic neuroscience of learning, and this means beginning not with humans, but with a strange creature known as a sea hare.

Don't poke the… hare

The sea hare is a type of sea slug. This large mollusc lives in shallow water and is named for two long protrusions on its head, which resemble (a little) the ears of a hare. These structures are actually used to detect chemicals dissolved in seawater, so the animal can 'smell' its way to food or a mate. Sea hares also have a set of gills on their back (which allows them to breathe) and a siphon which pumps water over the gills. Their siphon and gills are particularly delicate, so when the creature is disturbed it rapidly withdraws them into its body, like a snail retreating into its shell. It is this reflex which has formed the basis for a myriad of learning experiments by neuroscientists all over the world, and even won the Nobel Prize for Eric Kandel from Columbia University in 2000.

As well as this easily seen reflex, sea hares only have around 10,000 neurons, which are very large. This makes it relatively easy to study the circuits of neurons that control the animal's behaviour, especially learning and memory. And it turns out these simple creatures can, and do, develop memories. Normally, a gentle poke is enough to trigger the withdrawal reflex, but if you poke a sea hare *repeatedly*, after a while it will stop responding, having learnt that there is no real danger. This is a process called habituation and is arguably the simplest form of learning. If you have ever noticed a strong smell when first entering a room, only to realise a few minutes later that you can no longer smell it, you have experienced habituation for yourself. Basically, if the same stimulation is repeated over and over again, your brain realises it isn't important for your survival and stops responding to it. The same thing happens in humans and in sea hares.

To fully understand how this learning happens, we need to zoom in on those giant neurons in the sea hare's nervous system and look at how they communicate. When the animal is first prodded by the experimenter, this is detected by a sensory neuron. An electrical signal travels down this neuron until it reaches a synapse – the gap before the next neuron. Here, chemicals are released, which travel across the synapse and activate receptors on the second neuron. The most important chemical crossing these synapses is glutamate. This is an excitatory neurotransmitter, so when it binds to receptors it makes the second neuron more likely to fire. When enough receptors have been activated, an electrical signal is sent along the second neuron. The message has been transferred. This process is repeated until the signal reaches a motor neuron, causing the animal to pull in its siphon and gills. The reflex is complete. So far, so simple.

When the stimulation is repeated, the signal must be sent over and over again, and this is where learning can occur. When a neuron is first activated, it has plenty of glutamate to release, carrying the signal across the synapse to the second neuron, which can respond. But Kandel found that if he kept activating the first neuron faster than it could reabsorb or create more of these chemicals, it would eventually run out. The next time he activated it, there was nothing to carry the signal across the synapse, and the animal wouldn't respond. This 'neural fatigue' was a temporary effect, as glutamate levels can be restocked,* but if the repetitions are continued over a longer time-scale, the neurons undergo other changes, including the second neuron removing some of its receptors, making the synapse weaker. This is a much longer-lasting effect; in sea hares, habituation can last for weeks, suggesting a form of long-term memory.

Perhaps a more interesting phenomenon than habituation is that of sensitisation. Sensitisation can happen in lots of ways, and we will all have experienced some form of it. If you have ever watched a scary movie, then found yourself jumping at every little creak of the floorboards, you have felt it for yourself. Kandel found that his sea hares could be sensitised to a poke from a researcher by first giving them a small electric shock. This sensitisation made their response to the poke more dramatic. In the sea hare, it turns out this process relies on a chemical called serotonin.† The shock causes interneurons, positioned near the sensory neurons, to release serotonin. Serotonin binds to the sensory neuron, causing it to release more glutamate when activated. This

* This process isn't well understood, but one study found that replenishing glutamate in a mouse's neuron can take around 15 seconds at room temperature, although it was faster at the higher body temperature.

† Serotonin is better known for its involvement in mood, which we will cover in Chapter 4, but also has roles in a whole host of other brain processes, including learning and memory.

strengthens the synapse connecting the sensory neuron to the motor neuron, meaning a poke that would previously have been ignored now causes a dramatic reaction. Serotonin is also involved in classical conditioning, such as Pavlov's dogs' learnt association between a bell and food.*

So what does this have to do with learning in humans? We might not have siphons, gills or the ability to shoot ink out at people who are annoying us† but it turns out that on the inside, we are a lot like a sea slug. We have around 8.5 million times as many neurons, but using these to learn something new and remember it seems to involve a surprisingly similar process.

In 1949, a scientist called Donald Hebb put forward a theory often phrased as 'neurons that fire together wire together'. This means the more you activate the same pairs of neurons at the same time, the stronger the connection between them becomes. In a way, this is the opposite of the sea slug's habituation. Rather than the connections between neurons becoming weaker when activated, as in the case of long-term habituation, they become stronger, so when you next use them it is easier for the signal to flow. Whether a set of neurons undergoes this strengthening (known as long-term potentiation or LTP) or weakening (known as long-term depression or LTD) depends on a whole host of factors, including where in the brain the neurons are found, and the patterns of activation. But for simplicity, we will focus on LTP for the moment.

---

* Pavlov discovered that after repeatedly hearing a bell before they received their food, dogs could learn the relationship, and would begin salivating on hearing the bell. This is known as classical conditioning

† Sadly – that could really come in handy on the tube at rush hour!

Just like the sea hare, many of our synapses use glutamate to communicate. When a signal travels along a neuron to the synapse, glutamate is released, and travels across the synapse to activate the second neuron. If these signals are repeated, but with enough of a delay between them to avoid neural fatigue, the second neuron can be activated over and over again. This triggers changes in that neuron which free up more receptors. The first neuron also begins to release more glutamate in response to each signal. Both changes improve the chances of the second neuron receiving the signal and continuing to pass it on each time the first neuron fires. They also increase the speed of the transmission. This is why once you have learnt something thoroughly, it is much easier to recall it, and it feels like it takes less effort. It really is easier for your brain to activate these pathways, thanks to changes in how glutamate is released and received.

Long-term potentiation is given that name because, unlike neural fatigue, it can last for a very long time. The changes mentioned already can last for hours, but if activation of the pathways continues, especially if it is spread over a long time-period, the second neuron will build more receptors from scratch and add them permanently. This could be the way in which our brains store long-term memories forever. But it doesn't stop at the level of the neurotransmitters. If the pairs of neurons are stimulated together even more often, they release chemicals called growth factors, encouraging new synapses to form between the two neurons and further strengthening the connection. New connections develop, and these house your memories, theoretically for the rest of your life.

I remember it well...

The storage of memories for the long term is known as consolidation, and it is a fascinating process. Memories are first formed in the hippocampus deep within your brain, via long-term potentiation. They can remain here for hours or even days, but if every memory you ever formed stayed in your hippocampus forever, your memory would be limited. There are only so many neurons in this region, so they can only store a certain number of traces. Instead, the brain does something clever. Slowly, gradually, over days, weeks or even months, memories stop using the hippocampus, and instead are stored in the cortex, the surface layer of the brain. This highly folded layer is much larger, and contains a huge number of neurons. As each memory is stored as a combination of neurons, there is a vast capacity for storage here.

We don't know exactly how this process works, but there are two main ideas. After learning, the hippocampus repeatedly activates regions of the cortex, strengthening connections there, and these connections are where the memory is stored long term. Originally, scientists thought that the hippocampal memory formed first, and this process transferred it to the cortex, but recent research suggests the two memory traces might form at the same time. However, the cortical memory is, at first, unusable. Rather than transferring a memory from the hippocampus to the cortex, the reactivation may 'mature' the pre-existing cortical memory in some way, so it can be used. What is for sure is that the hippocampus is vital for this process and damage to this area makes it impossible to form new long-term memories.

If memory were like a video camera, the process would stop here. Memories have formed, been stored temporarily in one place (like on a memory card) then transferred elsewhere for long-term storage (like moving your recordings to a hard drive). When you wanted to recall the memory, all you would need to do is find it on the hard drive and 'watch' it back. But sadly, human brains don't work quite like that. When we recall a memory, we actually reactivate it, putting the neurons back into a state similar to when the memory first formed. And because the neurons are back in this flexible state, they can change, altering the memories they contain. When the memory is stored again, any new information is stored with it, and the next time you recall it, it is impossible to tell the original memory from the new parts. A false memory has been created.

Work on false memories was pioneered by Elizabeth Loftus, and her studies revolutionised our understanding of the way memory works. In the 1970s, she and her colleague John Palmer were interested in eyewitness testimony, and whether interview questions could change people's memories of a crime, so they devised an experiment. They asked students to watch videos of car accidents and describe what they had seen. Then they asked a series of specific questions, including the important one: 'How fast were the cars going when they ______ each other?' They filled the gap with different verbs, ranging from 'smashed' to 'contacted', and recorded the speeds the participants guessed. As the researchers expected, the more violent the word used, the faster the participants thought the cars had been travelling, an average difference of almost 10mph!

Changing a single word in a question seemed to affect people's memory of an event.

But Loftus and Palmer weren't satisfied; they wanted to see if this change persisted. So in a second experiment they showed another set of students a car-crash video, and divided them into three groups. They asked one group, 'How fast were the cars going when they hit each other?', the second, 'How fast were the cars going when they smashed into each other?' and the final, control group was not asked a question at all. A week later, the students returned and were asked to recall the video and answer some questions about it, including whether they had seen any broken glass (in fact, there was none). It turned out participants in the 'smashed' group were more likely to say they *had* seen broken glass. The word used really had tainted their memory of the event.

These studies and many others Loftus devised over the years tell us something important about our memory. It is far from infallible. In fact, it is fragile and easily changed. Something as simple as the way a question is asked can alter our memory of an event, and this change can persist, potentially for the rest of our lives. If you have ever had a disagreement with a friend or family member about the details of an event, it's worth bearing this in mind. Your memory may seem clear and accurate, and theirs obviously incorrect, but chances are you are both wrong! Discovering this makes me wonder how accurate my memory of my 18th birthday night out really is. Perhaps in telling people about it the next day, I inadvertently changed my memory of it. It is a little disconcerting to know that you can't rely on the accuracy of your own memories, but research in this area is starting to give us a fascinating insight into how our brains work.

But if our memories are that changeable, and inaccurate, why is it my memory of that birthday *seems* so clear? And do I *actually* remember it better than other days that year, or does it just feel that way? One reason it stands out so much is probably because it was a very emotional day. The happiness I felt when my friends presented me with the most beautiful pair of shoes I had ever seen, which they had clubbed together to buy me. The nervous excitement of using my ID for the first time. The anticipation of my party to come that weekend. All these emotions would have been impacting my brain, and changing the way the memories were stored.

Emotion and memories are inextricably intertwined. The amygdala (part of the brain's limbic system, which processes emotions) is found right next to the hippocampus, and for good reason. We have evolved to remember things that might be useful for our survival, so our brain is more likely to store information about something that causes a large emotional response, whether that is positive or negative.

As we will see in Chapter 4, when emotions are running high, our bodies release a whole host of chemicals, including adrenaline, noradrenaline and cortisol. It seems to be adrenaline that is responsible for the memory boost, as this chemical increases activity levels in the amygdala and hippocampus. Studies have found that blocking adrenaline receptors can reduce the emotional enhancement of memories and that increasing the amount of adrenaline released (with a drug or by asking volunteers to stick their hand in iced water) increases it. This is why we remember emotional events, whether negative like the death of a loved one or positive such as a wedding, more vividly than a normal day. And this is one of the reasons I remember my 18th birthday so well – it was a happy and exciting occasion, so my brain chemicals ensured I stored it for the future.

There is another type of emotional memory that can occur when the event is extremely dramatic, and this is known as a flashbulb memory. These are generally formed when there is a huge and shocking public event, such as the attacks on the twin towers in New York or the death of Princess Diana. Generally, people remember where they were when they heard the news in a much more vivid and visceral way than with most memories.* You might remember sensory information, such as the texture of the jumper you were wearing, or the colour of the TV reporter's hair. Whatever it is you remember, you will probably feel very confident that the memory is accurate, because it feels so vivid. But this is where things get interesting. When they are tested, these flashbulb memories are actually no more accurate than a normal memory. You may *feel* more confident about

---

* I have two flashbulb memories. One involves sitting cross-legged on the floor of our holiday cabin in the French countryside, seeing the news about Diana. I remember the wooden walls and floor, the smell of the trees, and knowing something really bad had happened, although I was too young to really understand it at the time. My second is from the 2005 London underground bombings. This one takes me back to breaktime at my senior school. I vividly remember the sight of girls filling the corridor, from the lockers on one side to the windows on the other, and the sound as we all desperately tried to get through to our parents on our phones to check they were safe.

them, but just like other memories they change and become less accurate over time.

Many people, for example, claim they found out about the twin tower attacks by seeing live television footage of the first plane hitting the towers. Even President George Bush claimed this when asked about that day. But there *was* no live footage of the first plane and videos only became available after the event. However vivid and accurate they feel, these memories have been created after the event. With our current understanding of memory formation in the brain it is easy to see how this happens. A few days after the event, maybe you saw a TV news report containing the footage, and this caused you to recall your memory of the day. Your neurons entered their flexible state, allowing the film clip you watched to alter them. The memory was then stored again, complete with the new information, changing it for the future.

Since Loftus's discoveries, scientists have been trying to work out what happens in the brain when we recall memories that makes them susceptible to these changes. While research is still ongoing, we now know that the process of reconsolidation (storing recalled memories again) is similar to the original consolidation that transferred them from temporary storage in the hippocampus to long-term storage in the cortex, but not identical. For both processes, the brain needs to build new proteins, but the exact proteins used are different. This leads to some interesting possibilities. If memories can be changed, could we selectively wipe out or alter, for example, traumatic memories in people with Post-Traumatic Stress Disorder (PTSD)?

Memory manipulation

PTSD is a condition that is triggered by a traumatic event, such as being held at gunpoint, or having a serious car accident. Survivors relive the experience repeatedly through flashbacks and nightmares, and feel panicky and on edge. They may also have problems sleeping, and feel numb or disconnected from the world. When people with PTSD have these flashbacks, they re-experience the emotions felt during the event, as strongly as they felt them originally. With most memories, over time, the emotions associated with them fade. We know that they are memories of an emotional event, but we don't *feel* the emotion so strongly when we think about them, especially if they happened a long time ago. But in PTSD, a glitch in the storage process means the emotion never fades. This can make it extremely difficult for survivors to function normally.

Amy Milton, Senior Lecturer in Psychology at the University of Cambridge, is working on understanding memories at the molecular level, and using this to help people with PTSD. But why are some people left with these kinds of memories after trauma, while others aren't? Milton explained to me:

> The hippocampus is very sensitive and pretty much shuts down under very high levels of stress. But the amygdala if anything ramps up its processing. So when somebody has been through a traumatic event, a particular set of predisposing factors is going to determine whether you tip the balance in favour of the amygdala over the hippocampus. And if that happens you end up with this fear memory that doesn't really have that time and date and context and

> space information that the hippocampus would normally provide. It's a fear memory that becomes very generalised and applies everywhere. So even when the patient is in a safe situation, these memories come back. Clinicians talk about working with patients who realise it will be something really innocuous like a red post box that triggers flashbacks of their bad car accident because the car that hit them was red.

Most fear memories are useful. If we are hit by a car because we were on our phone and stepped out into traffic, it is important we learn not to do that again. But if these memories are over-generalised, they stop being useful, and start being harmful. This seems to be what happens in people with PTSD.

Current treatments for PTSD are hit and miss, and extremely intensive. Patients have to recall their trauma over and over again, in a safe situation, in the hope that new memories will override the old traumatic ones. This probably works by the prefrontal cortex, the rational part at the front of the brain, learning to inhibit the amygdala's response, but is only successful in 50 per cent of people, and drop-out rates are high. So scientists are looking into whether using our understanding of reconsolidation, and drugs that can alter it, might help develop better treatments and improve patients' lives. Milton explained more:

> The idea is that you can actually get the memory back into this state where it can be unstable, and then either introduce new interfering information and structurally destroy that memory or give drugs. In an emotional learning situation, adrenaline is high, which facilitates the glutamate-based

> pathways that allow the memory to consolidate. So if you targeted that system and prevented it from working, you might be able to weaken that emotional memory.

Many of these studies, on humans at least, use a drug called propranolol. Used to treat high blood pressure, amongst other things, propranolol blocks one of the adrenaline receptors.* In the heart, this prevents adrenaline binding to heart muscle cells, reducing their activity and lowering heart rate and blood pressure. But propranolol can also cross the blood–brain barrier and affect the brain. Blocking adrenaline and noradrenaline receptors here is where it has an impact on emotional memories. If a memory becomes unstable, and there is propranolol in the brain, the effect of adrenaline is reduced and the memory may be reconsolidated with less of an emotional component. It may also directly block the creation of the proteins needed to consolidate a memory.

But it turns out that getting the memory back into the flexible state isn't as easy as was originally thought. As well as recalling the event, there has to be something unexpected involved in the recall for the memory to become flexible again, a 'prediction error'. This makes a lot of sense: you only want to update a memory if there was something wrong with the original one.† Milton

* The β-adrenergic receptor, hence it is one of a class of drugs known as beta-blockers.

† This might seem surprising in relation to Loftus's work, but a lot of the time she was probably inducing this 'prediction error' in her experiments without realising it. It may also be that prediction error is less important in the kinds of memories Loftus was interested in – episodic memories.

thinks that might be why there were several unsuccessful attempts to use propranolol to treat PTSD. The memory was simply recalled, so didn't become changeable. 'You need to introduce some kind of surprise, but it needs to be similar enough to what was learned before so that the individual doesn't think it's a completely different situation and just make a new memory.'

In the lab, this is done by training rats that a tone predicts a mild electric shock to their feet, which creates the fear memory. Then, the rats are put back in the box where they learnt this association, and hear the tone, but don't get shocked. This is enough to create a prediction error, and cause the fear memory to become unstable again. To see if this was important in humans too, Merel Kindt, Professor of Experimental Clinical Psychology at the University of Amsterdam, and colleagues devised a test that uses sensitisation, just like Kandel's experiment with sea hares.

In humans, one common form of sensitisation means that if you are afraid, you will jump more when a loud noise is played. To induce fear, Kindt taught undergraduate students that a picture of a particular spider was associated with a mild, but unpleasant, electric shock. When they saw that spider, they became jumpier, and reacted more to a loud noise, compared to when they saw a picture of a different spider. She found that if the students were given propranolol after seeing that spider again (with no shock this time), this extra jumpiness disappeared. But they needed both elements. Giving propranolol when students didn't see the spider had no effect, and neither did just seeing the spider in the absence of a shock. All of the students could remember, when asked, which spiders were linked to the shock; it was only their reactions that differed. So it seems that blocking reconsolidation after reactivating the fear memory removed the emotional response.

Despite these exciting discoveries, trials on people with PTSD are in the very early stages, but they do show promise. It may be that this combination of reactivating a memory, using a prediction error, and treating with propranolol at exactly the right time will turn out to be the perfect combination to help people with this devastating condition. But only time, and more studies, will tell.

Are you paying attention?

At this point, it would be easy to think that we understood the basic principles of learning and memory, and the role our brain chemicals play in them. Glutamate carries signals across synapses, and changes in how much is released and detected are the first step in allowing us to learn. Serotonin is also important, helping modulate the strength of synapses in the short term, before slower changes in receptor numbers or the growth of new synapses stores memories for the long term. Adrenaline gets involved too, telling us which memories are most important to store, because of their emotional content. But, of course, humans are a lot more complex than sea slugs, and the things we learn are more varied. This means there are other factors that can influence how we learn and remember, and our brain chemicals also play a part here.

If you have ever found yourself mid-conversation with a colleague who's telling a long and boring story and realised you haven't got a clue what they have been saying for the last few minutes, you may have fallen foul of acetylcholine. This brain chemical is involved in alertness, attention and motivation, something that may be lacking when listening to anecdotes about Gary's new lawnmower! At a more fundamental level, acetylcholine helps you switch your focus between external stimuli (such as your work colleague) and internal (like what you are going to have for dinner). This is incredibly important, not just for learning new information, but for storing and retrieving it as well.

Circuits in your brain can be divided into two types: those that provide information about the external world, and those that process that information internally. Often the same neuron will receive input from both external and internal circuits, so it has to 'decide' which is more important at any point in time. This is where acetylcholine comes in. When levels of this chemical are high, it boosts your response to external input and suppresses internal feedback. This puts you in the perfect state to learn new things, as you are focused on taking in new information. After a while, however, acetylcholine levels drop, and your attention starts to drift. During slow-wave sleep (the deepest phase of sleep, when brainwaves sync up and slow down, see chapter 5) and when you are awake but resting, acetylcholine levels in the hippocampus are lower, more glutamate is released by internal circuits and these become dominant. This allows memories formed during focused wakefulness to be stored as long-term memories.

Acetylcholine also helps with memory recall. When trying to retrieve a stored memory, your brain activates *other* concepts relating to that fact. For example, I have been trying to teach myself Spanish recently, but have found that the French I learnt as a child (and which I thought I had forgotten entirely) is holding me back. Quite regularly, when I try to retrieve a Spanish word, the French one comes to mind, so I end up with sentences like 'Hola, je m'appelle Ginny.' While frustrating, this makes a lot of sense once you know how the brain stores memories. We don't store information exactly as it is; rather we extract the gist from it. So rather than remembering every word in this chapter, your brain will (hopefully) process the information it contains and produce a nice, concise, take-away message for you.

But this is where problems can start. My brain has stored a link between the phrase 'my name is' and 'je m'appelle'. Now, I am trying to learn 'me llamo', but every time I go to store that, my brain could also activate the old, French trace. Because of the way memories work, the more a trace is activated, the more efficient it becomes, as the synapses strengthen. So my repetition of 'my name is' = 'me llamo' might be enhancing the Spanish trace, but it is also strengthening the French one. So how can I ever hope to be able to speak Spanish without peppering my conversation with French words? Acetylcholine to the rescue! When brain levels of this chemical are high, attention is focused on the external world, and internal pathways become weaker. This means I can activate the 'my name is' = 'me llamo' memory trace, strengthening it without reinforcing the French version as well.

We know a lot of this because of experiments using drugs that change the levels of acetylcholine in the brain. One widely studied drug is scopolamine, which is used to treat motion sickness. Scopolamine blocks acetylcholine receptors, meaning the brain no longer responds to the chemical when it is released, tricking the brain into thinking levels are low. When people are given this drug straight after learning a list, they are able to recall the words without any problems. But if an injection is given *before* the learning session, their recall suffers. Unable to detect the acetylcholine their brains are producing, the subjects get stuck in internal mode, happily consolidating old memories but finding it hard to form new ones.

Interestingly, these drugs aren't a new discovery made by researchers working in neuroscience – quite the opposite. Jimson weed is a plant in the nightshade family (which also includes potatoes, tomatoes and tobacco) which was used as an ancient medicine and to anaesthetise people during surgery. Its hallucinogenic properties have also led to its use during religious rituals and recreationally for centuries. Its active ingredients include scopolamine, and this is thought to be responsible for the severe amnesia that users experience.*

In fact, some scientists believe that scopolamine is to blame for the hallucinogenic effects of Jimson weed as well. With the brain unable to detect acetylcholine, this may cause a dramatic shift to focus on internal perceptions. This prioritisation of consolidation and recall of memory rather than storage could mean that the drug user starts to 'experience' their memories, believing them to be real and current. Effects are usually temporary, although

* Which makes me wonder whether Jimson weed *actually* anaesthetised patients, or whether they just couldn't remember afterwards the excruciating pain they felt during surgery.

memories of the time when the drug was in a person's system rarely return, fitting with the idea they weren't stored in the first place. Sadly, the plant can be fatal, in doses not much higher than needed to produce the hallucinogenic experiences.

Limitless?

If there are drugs that can block memories from forming, are there any that we could use to *improve* our memory, and help us learn faster and more efficiently? Despite studying at Cambridge not *that* long ago, and despite the surveys that show how common these 'study aids' are at universities, personally I have never taken one. But so-called 'smart drugs' are a rapidly growing area and claim to do just that. Many commonly used 'smart drugs' are available on prescription. Adderall and Ritalin, for example, are both stimulants used to treat ADHD. But for many years, these pills have been used off-label by students and others looking for an edge. Both boost the levels of brain chemicals dopamine and noradrenaline, which is beneficial for people with ADHD as these systems are thought to malfunction in the condition. At the low doses taken by patients, this boosting seems only to occur in the frontal regions of the brain, which are involved in forward planning and carrying out complex tasks. Interestingly, take too high a dose and it can cause negative effects on cognition, as the drug starts to affect other brain regions as well.

In people without ADHD, boosting levels of these neurotransmitters just enough is thought to improve concentration and make boring tasks more enjoyable. No wonder it is used as a study aid! But are they really having the dramatic effect people claim? Recent analyses of many studies into these 'smart drugs' have found small effects on cognition, problem-solving and memory in healthy people, but only in some people and on some tasks, and these seem to vary depending on the study. There is also a suggestion that compared to a placebo, the drugs improve confidence in one's own abilities, so people rate them as having more of an effect on their brainpower than they actually have. As they are stimulants, there is also a risk of abuse. Taking a large amount of the drug boosts dopamine more dramatically, meaning people can become dependent on them (as we will see in Chapter 3). Recreational users of these drugs may come to crave the 'rush' they provide, and feel unable to function without their 'little helper'. For this reason, it is unlikely these drugs will become available for general use as a brain booster any time soon.

Another drug that has received a lot of interest recently, and which seems to have fewer side effects, is modafinil. Originally developed to reduce daytime sleepiness in people with narcolepsy, it has the ability to improve cognition even in people who are extremely sleep-deprived. Because of this, a whole range of potential users have been suggested, from the military to surgeons to astronauts! And, of course, there are others who are keen to try out this 'wonder drug' and get a boost in school or in work. Modafinil's mechanism of action isn't well understood, but it affects the levels of a whole range of neurotransmitters including dopamine, noradrenaline, serotonin and glutamate. How exactly this boosts memory and cognitive control remains to be seen, but modafinil's growing use led neuroscientist

Barbara Sahakian, who works on these drugs, to argue recently that they should be licenced to allow proper control over them.

As well as the scientific questions these drugs raise, there are also ethical questions to be discussed. Is taking a brain-boosting drug before an exam cheating? What if you only take it to study? Many of us use caffeine to help us stay alert and attentive when we have a tough day* and caffeine is a brain-altering drug too (see Chapter 5). So how is that different? Sahakian has argued that modafinil is actually preferable to caffeine in many situations because it doesn't cause side effects like hand tremors. For surgeons about to undertake a long procedure, for example, a wakefulness-promoting agent that doesn't give them the shakes could be a huge benefit to everyone involved.

But if it suddenly becomes the norm for everyone to dose up, might it lead to a world where you *have* to be on drugs in order to compete? That certainly isn't a future I aspire to. And it could make things even more difficult for children from less-privileged backgrounds. How wealthy your parents are already has a huge effect on your future, for a range of reasons, from more time spent being read to as a child to the availability of extra tutoring if you struggle with a topic. But if you add to that expensive brain-boosting drugs, only available to those who can afford them, the divide could widen even further.

So, would I try these drugs? I'm certainly tempted. The benefits of modafinil are intriguing, and its

---

* I know this book wouldn't have been possible without a strong cup of coffee or two!

short-term safety seems pretty well tested. But I would need to be sure I was getting the real thing. Currently most people using these drugs off-label are ordering them from websites, which means there is no guarantee they are getting what they paid for. It could be anything in those pills, and that's a risk I'm just not willing to take.

There also isn't a lot of long-term safety data available yet. While a single use or two is unlikely to cause any harm, we know that the brain adapts over the long term to anything we take. That's why your morning coffee doesn't provide the jolt it once used to, and you may find yourself reaching for another double espresso (see Chapter 5). And it's why drug users constantly have to increase the amount they take to get the same hit (see Chapter 3). So, if you were to take modafinil regularly, would it keep having the same effect, or would you begin to need it to work at a normal level, becoming as dependent on it as we are on caffeine? These are all unanswered questions that need to be resolved.

We also have no idea how it would affect developing brains, so the growing use in schools and universities is of particular concern. Babies' and children's brains are in a different state from adults' brains, as they rapidly change and learn. In fact, our brains don't finish maturing until our mid-twenties. And we are still working out exactly what is going on with our brain chemistry at each point in this process, so the idea of altering it with unnecessary drugs is, to me at least, very worrying.

Babies and chicks

Babies and young children have evolved to learn. Compared to most other animals, our infants are born incredibly helpless. Thanks to our big brains and upright stance, they have to be born before they are really ready, and because of this, they are entirely reliant on their caregiver. They can't walk, feed themselves or even regulate their own temperature. This leaves them vulnerable, but also means they can be shaped by their environment in a much more dramatic way.

Infants are born with about as many neurons as adults, but that doesn't mean their brains are fully formed – far from it. As we have seen, it is the connections between those neurons that are important for learning and memory, and these change rapidly over the first few years of life. During these years, our brains are in the perfect state to form new connections via long-term potentiation. But the perfect time for this process varies depending on what part of the brain, and what ability, you are looking at. These 'windows of opportunity' are known as sensitive periods, and a lot of what we know about them started with an Austrian researcher called Konrad Lorenz.

Lorenz had always loved animals. As a child he kept a host of pets, and even helped a nearby zoo to look after their sick charges. But his father, a surgeon, was adamant he should study medicine, so in 1922 Lorenz began his studies. Despite this, Lorenz couldn't give up his work on animals. He kept diaries, documenting his observations, and in 1927 his diary about a jackdaw was published in a prestigious journal. After graduating with his medical degree, Lorenz completed a PhD in zoology.

In the 1930s, following in the footsteps of his mentor Oskar Heinroth, Lorenz began working on greylag geese, and a phenomenon he later called 'imprinting'. When chicks hatch, many species need to rapidly learn to follow their mother, wherever she leads, to avoid getting lost. Heinroth noticed that the goslings of greylag geese would, if the conditions were right, begin following a human, rather than their mother. So his student Lorenz set about finding out how this happens.

Through various experiments, Lorenz discovered that the goslings are born ready to form an attachment to the first large, conspicuous moving object they see. In the wild, this would be the mother, but in the lab the birds would imprint on models of other species, a ball, or even (most famously) Lorenz himself. In one experiment, Lorenz took half the eggs from a nest, and raised them in an incubator, while the other half were raised by the mother goose. When hatched, as expected, his chicks imprinted on him, while the others imprinted on their mother. Next, Lorenz mixed up the chicks and covered them with a box. As soon as he removed the box, his chicks ran to him, while the others headed straight for their mother. He argued that this imprinting must happen during a critical period, just after the chick has hatched. If the gosling saw no suitable stimulus until after this point, it would never imprint. He also believed imprinting was permanent, and would affect the animal for the rest of its life.

His career was interrupted by the war and in 1941 he joined the German army as a medic, later becoming a psychologist in an SS unit. While he later denied involvement, or knowledge of the atrocities they committed, it seems that Lorenz used his work to support Nazi ideology. As well as looking at imprinting in chicks, he was interested in how this affected later sexual behaviour, and discovered that a duck raised by geese would later be sexually attracted to geese.

He used this to argue that hybridisation (the mating of two individuals of different species) in animals would cause confusion because their innate drives might conflict, and they might end up 'weaker' because of this. Of course, this fitted in perfectly with Nazi ideals of 'racial purity' and preventing interbreeding, which might 'weaken' the human race. He also believed that domesticating animals made them weaker, and was 'afraid' of what this might mean for humans. In the biography he wrote when receiving the Nobel Prize in 1973, he said:

> I wrote about the dangers of domestication and, in order to be understood, I couched my writing in the worst of Nazi-terminology. I do not want to extenuate this action. I did, indeed, believe that some good might come of the new rulers. The precedent narrow-minded Catholic regime in Austria induced better and more intelligent men than I was to cherish this naive hope. Practically all my friends and teachers did so, including my own father who certainly was a kindly and humane man. None of us as much as suspected that the word 'selection', when used by these rulers, meant murder. I regret those writings not so much for the undeniable discredit they reflect on my person as for their effect of hampering the future recognition of the dangers of domestication.

Lorenz continued to work for the Nazis until he was captured by the Russians in 1944. There, he was put to work as a doctor, until he finally made it home in 1948, and returned to his research. There is no doubt that Lorenz was an important figure in our understanding of animal behaviour and he is often referred to as 'The father of ethology [the study of animal behaviour]'. But science doesn't occur in a vacuum, and it isn't helpful or even possible to separate someone's scientific beliefs from their political ones. A scientist's views about the world will drive the questions they ask in their studies, the way they interpret their results, and the inferences they make from them. This doesn't mean we should write off the discoveries Lorenz made, just that we must look at them through a lens of the political issues of the time.

In the 1950s, Lorenz's work was picked up by a German-born US researcher called Eckhard Hess. In his lab in Maryland, he investigated ducklings, finding their imprinting occurred between 13 and 16 hours after hatching. This led to renewed interest in the area, and over the following decade many more discoveries were made. It turned out that while Lorenz was right about the window of opportunity, it wasn't as fixed as he had thought. If chicks were kept in social isolation after birth, for example, their imprinting window stretched to 20 hours. So, imprinting doesn't just follow a set path, decided by biology. It can be affected by experience as well.

Now, researchers tend to refer to sensitive periods, rather than critical periods. These are times during which an animal is predisposed to learn certain things, and when their brain and future behaviour can be shaped by experience. And it seems that, like geese, humans too have sensitive periods for certain abilities.

Take vision, for example. When they are born, infants can't see in much detail.* Their peripheral vision is poor,

* Their 'visual acuity', measured by how fine a black and white stripe they can tell apart from a grey screen, is 40 times worse than an adult with normal sight.

and so is their ability to detect colour and contrast, but these abilities rapidly develop. To do this, though, the eyes need to be working properly. This means babies born with cataracts can have lifelong problems. Cataracts, when severe, block all visual information except for a general sense of lightness and darkness. When they are removed, the eyes can function normally, in theory at least. In adults, how long the cataract has been in place doesn't seem to matter much, in that once it is removed, vision is restored. But in babies, the story is very different. If the cataract is removed in the first six weeks of life,* and suitable follow-up is done, most children will grow to have normal vision. The later the removal is left, however, the more visual problems the child will have and these will remain throughout their life.

Another clear example is language. Anyone who has tried to learn a language as an adult knows how difficult it is, but children do it, apparently, with very little effort. Move a young child to a foreign country and within months they will be speaking like a local, while their parents are still struggling with the basics, and likely will never lose their accent.

It's clear that timing is crucial for learning, and input early on is vital for an ability to develop properly. It is also clear that this window closes, and learning can become more difficult, or even impossible, afterwards. Partly, this is down to the way in which our brain develops. When a baby is born, its brain is rapidly forming new connections, up to one million every second. Then,

---

* For cataracts in one eye. Interestingly, if there are cataracts in both eyes, the timescale is slightly longer: eight weeks from birth.

as it grows, areas begin to mature. Starting with simple abilities like vision, connections that are used begin to strengthen via long-term potentiation, and those that aren't used are pruned away.

For example, newborn babies can distinguish between the sounds of any language. But as they are exposed to one language over others, they start to lose this flexibility. By the age of one, a baby exposed only to English will begin losing the ability to distinguish sounds that aren't used in the English language.

If you miss the age at which the neurons are in 'connection' mode, you miss the sensitive period, and input can't shape the brain in the same way. The pattern of development, and the switch from connecting to pruning, varies dramatically for different areas of the brain, and for different abilities. The frontal areas of the brain, involved in decision-making and deliberation, for example, don't finish developing until we are into our early twenties.

Scientists don't know exactly what it is that switches a region of the brain from one mode to another, but there are several theories. One suggests it might be the balance of inhibitory neurons, which use chemicals like GABA (gamma aminobutyric acid), to excitatory ones, which use glutamate. If there is less inhibition in the developing brain, this might make it easier for new connections to form. Neural competition may play a role too, with active neurons shutting down the less active ones around them. Experience is also important though. Depriving an animal of all sensory input seems to delay the closing of their sensitive period, at least for a little while. While there is still much to discover, the hope is that if scientists can work out how to reopen the sensitive period, they may be able to help people like those children whose cataracts weren't removed in time.

Ceauşescu's children

So it seems we do have sensitive periods for perceptual abilities. But what about more complex behaviours? Probably one of the most important things children have to learn is how to navigate the social world we live in – how to form bonds with people, and the rules of social interaction. And for most children, this happens naturally, as they soak up the behaviour of their caregivers, interacting both with them and with others around them. But what happens when this social interaction is taken away entirely? Sadly, events in the 1980s and 1990s have given us a pretty good idea.

In 1967, Communist Party leader Nicolae Ceauşescu became president of Romania. In his quest to make Romania a 'world power', he brought in a number of totalitarian rules. He believed that one way to achieve his aim was to increase the population of the country, so he banned contraception and abortion. This led to huge numbers of babies that parents were unable to take care of, so institutions were set up to care for them. They are often called 'Romanian Orphanages', but it is estimated 60 per cent of the children in them still had living parents; parents who were too poor to care for them, or who had been convinced they would have a better life being looked after by the state. They were even told they could return to collect their children later, if their situation improved.

But these institutions were not what they were made out to be. Quickly overloaded by the numbers of children being left there, the employees were unable to provide more than the most basic of care. In many, food and heating were scarce, and the children were abused both by their caregivers and by the older children. Many died from minor illnesses or starvation. At best, children were fed and kept alive, but they weren't held, or talked to, or played with. And while these things might not sound as important, they are vital for a child's social development.

When Ceauşescu was overthrown in 1989, the world became aware of these children, and the conditions they had had to endure. Charities were set up to help improve conditions,* and a large number of children were adopted by families around the world. Despite this, the numbers of children in institutions remained high, and more continued to arrive as the political situation worsened poverty in the area. Sadly, while they may have received more toys and supplies, the amount of care given to these children didn't improve much in the following decade.

As word spread, news of these children's plight raised the interest of researchers studying child development. Perhaps this dreadful tragedy could provide important insights into how neglect, both physical and social, affects the developing brain, and whether experiences later in life could make up for this. And maybe that could help other children in the future.

In 2000, researchers including Nathan Fox set up the Bucharest Early Intervention Project to try to determine

---

* I remember, as a child, packing shoeboxes full of toys and treats each Christmas, which were to be sent to these children. At the time, though, I had no idea of the extent of their misfortune.

just this. The project assessed over 100 children who had been abandoned around birth and placed in one of six state-run institutions in Bucharest, Romania. Then the researchers did something that must have broken their hearts. They randomised the children in to one of two groups. One group remained in the institution, while the other was fostered by a family, who were given special training for the study, and supported financially.* The researchers then followed these two groups, as well as a group of similarly aged children who had never been put into care, known as the 'community controls'. They found, as might be expected, that the institutionalised groups were behind in their development in a whole host of ways. They were behind on language skills and motor development, and their IQ was lower. They showed strange behaviour when around strangers, running to them and hugging them as if they were long-lost relatives, not showing shyness or wariness or turning to their caregiver for reassurance as most children will. When their brain activity was tested using EEG, it was generally lower. Their brains just weren't developing as they should be.

But it wasn't all bad news. As the researchers followed up over the next 12 years, they began to see a pattern emerge as the children they had removed from the institution began catching up. It seemed they were able to form a secure bond with their new caregivers, but

* This might sound unethical, but it is worth keeping in mind that had the study not taken place, all the children would have remained in the orphanage. Because of their work, a number of children were adopted into families who were provided with financial and emotional support, and the rest were in no worse a position than they had been in before.

they still lagged behind the group who had never left their families.

At the start of the study, the children ranged between six months and nearly three years old, so this provided the researchers with a unique opportunity. They were able to break down the fostered group to see if there were any differences based on how old they were when they had been taken out of the institution. This might tell us whether humans have sensitive periods for social development. And they found something important. The younger the children were when fostered, the better they were able to catch up. But the exact age differed for different abilities. Children fostered before 15 months of age showed rapid catch-up in their language abilities, while after this they had more trouble. For IQ, the vital age was 18–24 months. At eight years, children fostered before they were two showed normal EEG activity, while those fostered later still showed differences.

This suggests that there are sensitive periods for human infants, during which they need the right kind of stimulation to develop. As in animals, it seems the basic abilities, like sensory and motor skills, develop first, with more complex social behaviours having their sensitive periods a little later. But why? What is going on in the brain of neglected babies that can affect their behaviour for years to come?

Cry baby

Researchers believe that some of the most critical experiences for infants are 'serve and return' interactions. This can be as simple as a parent talking to a baby, and responding to noises or facial expressions. As the infant grows, it might include shared experiences like pointing at objects, tickling or playing. The important thing is that the caregiver is reactive to the baby. In one experiment, scientists asked mothers to look at their baby with a blank expression on their face, and not to react to them in any way. Within seconds, the infants became visibly distressed. We think that they take this lack of reaction as a sign of danger. This makes sense; for a baby entirely reliant on a caregiver for survival, lack of response could be deadly.

So this lack of interaction activates their stress response, flooding their body with adrenaline and cortisol. If this happens occasionally, for example when a baby is left to cry while a normally responsive caregiver is busy, this doesn't do any long-term harm. In fact the baby might learn to self-soothe. But if the stress system is consistently activated for long periods, it can cause all sorts of problems.

Consistently high levels of cortisol can damp down the growth of dendrites (the branching structures at the end of the neuron that allow it to connect to many others), and the formation of synapses. Considering how rapidly it changes during the first few years of life, it is not surprising a baby's brain is particularly sensitive during this time. Stress also inhibits myelination, the process by which the axons (nerve fibres) of the neurons, which carry messages around the brain, are covered in an insulating coating, often called white matter. This is a vital part of the maturation process, helping the neurons within the brain become more efficient.

Brain-scanning studies have shown that the children in the Romanian orphanages had smaller brains, with less of the grey matter that is made up of neuron bodies, and less of the white matter connections. Moving them to foster care helped their white matter recover, to some extent, but their grey matter volume didn't catch up. This fits with what we know about brain development, as myelination of a region comes after the time when the neurons are growing and connecting most rapidly. In fact, myelination doesn't fully finish in the brain until we are in our twenties.

This high level of cortisol also changes the way the chemical itself functions in the body long term. We all have a cycle of cortisol. It is highest in the mornings* and dips towards the evening. But children who have been neglected show blunted rhythms, which stay low throughout the day. This seems to link with physical growth issues, and also with indiscriminate friendliness, just as was seen in the Romanian orphans. But, just like brain structure, the cortisol cycle can change, and recovery has been seen in children who are fostered or adopted into loving families, as long as the caregivers aren't too stressed themselves.

This may seem a little confusing on the surface. Earlier we discussed the fact that emotions, and the brain chemicals involved, can enhance memories. But in the brain, while a little of something can be good, too much can have negative impacts, and this applies to cortisol, as well as other stress chemicals. It turns out the release of cortisol does help make certain areas of the brain more susceptible to long-term potentiation, helping with

---

* Interestingly, cortisol also increases your blood pressure, so its daily peak in the morning is one of the reasons heart attacks are most common early in the day.

memory storage. But cortisol can hang around in the bloodstream for a few hours and it only has this effect initially. Scientists think that it actually has two effects: a fast one, which helps with memory storage, and a slow one, which prevents it.

This makes sense from an evolutionary perspective. If something scary or stressful happens, you want to store that memory, so you can avoid it in the future. The adrenaline, noradrenaline and cortisol released help you do just that. But after that memory has been stored, and the threat has passed, cortisol sticks around, and prevents new memories from being formed over the top of that important one. It protects it, and ensures it is stored securely for the future.

The problem comes when cortisol release is ongoing, because of chronic stress. In this situation, you are always in the 'post-fear' mode, and long-term potentiation, and therefore learning, is constantly damped down. The less reactive cortisol system found in children from the orphanages may also mean that they don't ever experience the short-term benefits that this chemical can have on memory.

When it comes to our brain chemicals, it is all about balance. While it may be true that a chemical is involved in learning, that doesn't mean that increasing levels will necessarily increase your brainpower. Our brains are easily overloaded, and are looking for that 'Goldilocks point' with just the right amount of each chemical for them to function properly. This means we need to be cautious using drugs which change levels of these chemicals, particularly in healthy people. As we have seen, they can often have unintended consequences.

But as we discover more about the neuroscience of learning and memory, and the chemicals that control it, the hope is that we *can* find ways to make it more efficient, safely. This might mean better ways of teaching in schools, or drugs that can help children who have missed a developmental window because of disease or abuse. But, as with any development in science, there are ethical issues too. Improving treatment of PTSD or other memory problems could be life-changing for millions of people, but the risks if we find ways of tampering with memory are obvious, and take us into the darkest realms of science fiction and spy thrillers. For the moment, when it comes to boosting brainpower, the safest and most reliable methods aren't the newest, or the most exciting: practise, practise, practise, and, as we will see in Chapter 5, make sure you get enough sleep…

CHAPTER THREE

Getting Motivated

I put down the phone, my excited smile beginning to hurt my cheeks. My blood was pumping and I felt like jumping around the room. I couldn't wait to tell my partner, Jamie, what had just happened, so I sent him a text straight away. 'OMG – you will never guess who just called me! A well-known publishing company* wants me to work on a book with them! I had so many of their books as a kid – I can't believe it!!!'

As you can probably tell from the number of exclamation marks I used, I was pretty excited. Actually that's an understatement; I was bouncing off the walls. I had loved their beautifully illustrated books when I was growing up, and the idea of writing for them felt like a dream come true. At each milestone during that first book, the thrill continued. When I had the first pages back from the designers, when I proofread the final copy and then that moment when I held the book in my hand for the first time – each of these gave me a surge of pleasure, pride and a feeling of reward. When they got in touch about the follow-up, I was pleased. The next time, I didn't feel much of an emotional reaction at all, and by the time I got to the fourth one – yawn! I still enjoyed

* I thought I'd leave it to you to guess the actual company!

writing the books, but it was no longer something new, so didn't feel as thrilling.

Our brains have evolved to help us succeed in the world. To do that, at the simplest level, we need to eat and drink, stay away from danger and reproduce to pass our genes on to the next generation. So how do we know to do these things? Partly, it is down to a system in our brains, which rewards us when we do something that is beneficial for our survival, motivating us to do that thing again in the future. This system is the reason doughnuts taste so good when you are hungry, and the reason people will go to great lengths to impress the person they are dating and (hopefully) get them into bed.

In the modern world, the reward system motivates us to pursue one reward after another. A bigger house. A better job. More beautiful shoes you never wear. But the payoff once you reach your goal is short-lived, so you find yourself striving for the next. It makes sense evolutionarily to have a system which keeps us moving forwards, seeking the next reward. Once you have eaten that berry or slept with that attractive stranger, it is logical to move on to the next opportunity.

This was what happened for me with these books. Once I had achieved that goal, achieving it again didn't produce the same reward. This drives me to keep moving forward in my career and to keep looking for that next big achievement. I had another rush when I was contacted about writing this for Sigma, my first solo book, and I look forward to the excitement (and terror!) of releasing it to the world. So, this fleeting aspect of reward can have its benefits, but unfortunately it can also be our downfall.

One problem is that this system is a simple one, and it isn't always suited to the modern world. It doesn't understand that in our world of plenty, eating another cupcake is likely to be detrimental rather than beneficial to our health. So, often, we find ourselves drawn towards unhealthy behaviours.

The most extreme example of this system proving problematic comes from drug use. Drugs of abuse can co-opt this system, producing feelings more intense than any 'natural' reward. And this makes them very hard to give up. To understand how this system of drives and desires works in all of us, it makes sense to start by looking at drugs, and how they can (sometimes) lead to addiction.

Understanding addiction

Not everyone who tries a drug becomes addicted. Some drugs are more addictive than others, but there also seem to be some people who are more likely to become addicted. This suggests there must be something about their brains which interacts with the effect the drug has to put them more at risk. But to understand how and why this happens, we first have to look at what addiction really is.

Addiction is one of those funny words. We all feel we know what it means, but defining it is surprisingly difficult. We use it in ways that don't fit the clinical definition, saying we are 'addicted' to a new TV show, or a type of food. And while some would argue it is possible to be addicted to food (see Chapter 6), most of us aren't, because to be defined as an addiction, your desire for that item must become so great that you will lose track of other things in your life, pursuing it to the detriment of work, relationships and even your own safety. So what is it that drives people to these extremes?

One of the classical theories is that people who are addicted are in search of a high. They like the feeling of taking a drug so much that it drives them to pursue it, against all odds. But what makes drugs pleasurable? If you google this question most of the resulting webpages will tell you that the 'high' is due, at least in part, to the release of huge amounts of the brain chemical dopamine. Dopamine is a small molecule which was only discovered in the brain in 1957 by Kathleen Montagu.* It was initially known to be important in the initiation of movement, thanks to research on people with Parkinson's disease. This condition is characterised by the death of neurons in part of the brain called the substantia nigra.† As these neurons produce a lot of the dopamine in the brain of a healthy person, people with Parkinson's have low levels of dopamine. This leads to symptoms including difficulty with movements and balance, and tremors, usually in the hands. We now know that dopamine is important for drugs of abuse too.

---

* In many places, this credit is given to Arvid Carlsson, who was working on dopamine at the same time as Montagu and went on to win the Nobel Prize in 2000 for his role in its discovery. However, his paper, confirming that dopamine is a neurotransmitter in the human brain, was published several months after Montagu's. Sadly, Montagu died in 1966, and the Nobel Prize is only awarded to living scientists, so we can't know whether the honour would have been given to her alongside Carlsson.

† This literally translates as 'black stuff' – you can probably guess what this area of the brain looks like!

Take cocaine, for example.* Normally, when dopamine is released into a synapse, it diffuses across to bind with receptors on the second neuron, changing how likely it is to send a signal. Meanwhile, any excess dopamine in the synapse is sucked back up by the first neuron, where it is recycled. This process is known as reuptake and is controlled by dopamine transporters in the membrane of the first neuron, which you can think of as the vacuum cleaners of the synapse.

When someone takes cocaine, the drug enters the brain and blocks the dopamine transporters. This means that when dopamine is released into the synapse, it hangs around for longer, and at higher levels, than it normally would. This creates a feeling of euphoria, for a short amount of time, or so every website about cocaine tells you. But is it really that simple? And even if it were, what is it about dopamine that makes it the 'pleasure chemical' it is often touted to be? We can trace the beginnings of this theory back to one very special rat, and a mistake made by a pair of scientists.

The odd rat out

Probing the brain to understand what certain areas do and how they communicate is a challenge, but techniques have been developed and refined over the years to allow us to do just this, at least in animals. For example, brains use electrical signals to communicate, so if we apply electricity from an external source, we can activate specific areas. One way to do this is to perform an operation on a rat to implant a tiny electrode in an area of its brain.

---

* That's a figure of speech, and not a recommendation!

Scientists can then watch its behaviour when that area is stimulated to try to deduce what that area does. For example, if the rat's leg twitches when the electrode is activated, we know we have found a motor region.

This was exactly the technique used by Peter Milner and James Olds. In 1953, Milner was a graduate student working in the lab of psychologist Donald Hebb* at McGill University, Canada, when Olds joined the lab. Olds had a background in psychology from Harvard's Social Relationship department, but had never worked with animals, so Milner took him under his wing. Previously, Milner had been working on his PhD, but Old's work on motivation and learning sparked his curiosity, so he changed his plans so they could work together.

The pair implanted an electrode in a part of the rat's brain known as the reticular arousal system, which others had found was a kind of 'punishment centre'; rats dislike stimulation to this region, and will work to avoid the electrode being activated. They then allowed the rat to roam around, and stimulated its brain when it was in a certain place. To their surprise, they quickly noticed that the rat would return to wherever it had been when the electrode was stimulated. This shouldn't have happened; if the rat disliked the sensation, why would it return to the place it experienced it?

As every good scientist should, the pair decided to repeat the experiment with a different rat. But the results didn't match. So they tried again. Still no luck. When multiple attempts to repeat the experiment had failed, the pair began to wonder whether they had missed the

---

* The same Hebb whose work on learning we met in Chapter 2.

mark when inserting the initial rat's electrode, and a quick X-ray proved that this was the case. The electrode had ended up in a region deep within the brain they called the septal area. So, they set about repeating the experiment with more rats, implanting the electrodes in this new area of interest.

When they published their research in 1954, it was a huge step forward for neuroscience. For the first time, researchers had discovered a region of the brain that was intrinsically rewarding. They experimented by allowing a rat to control its own electrical stimulation, by pressing a lever, and found that when the electrode was in the septal area, the rat would continue pressing for as long as it was allowed, never appearing sated. It just couldn't get enough of this feeling. They began to wonder whether animals are driven to eat not just to stop the feeling of hunger, but because it activates this area, which is pleasurable in itself.

Over the next few years, while Milner went back to his doctoral studies, Olds continued this research, working with his wife and fellow neuroscientist Marianne Olds to map the brain for reward and punishment sites, and with a grad student to explore the relationship between food and brain stimulation.

Amazingly, follow-up experiments showed that direct activation of the reward areas was stronger than any natural reward that the team had tested. In one study they gave rats that hadn't had food for 24 hours the option of pressing one lever for brain stimulation or another for food. Again and again the rats would choose the brain stimulation. They were so obsessed with it that the researchers were convinced they would die of starvation unless they intervened. In some experiments, the animals pressed the lever 2,000 times an hour.*

In another study, the rats were placed in a box with a lever at each end, and an electrified grid on the floor between the two. This time, the rat could press the first lever three times to get their brain stimulation, but then had to run to the other lever – continuing to press the same one repeatedly had no effect. This meant they had to run between the two, over the grid, which could be set to give them an electric shock of varying strengths. Amazingly, the rats would brave the unpleasant shock to their feet to administer the brain stimulation. As a comparison, the researchers tested how large a shock a starving rat would bear to access food and found that rats would endure a bigger shock for brain stimulation than for food. The researchers had discovered something huge: they had found the areas of the brain that are responsible for feelings of reward or pleasure.

This pioneering research sparked a flurry of interest in the brain's reward system, and we are now starting to understand how it works and the brain regions involved. The area Milner and Olds were interested in, the septal area, is now known to contain the nucleus accumbens, which is one of the most important areas for reward. We now know there are other regions that are important too, so most neuroscientists talk about a 'reward circuit', rather than a reward area.

When it comes to brain chemicals, the most important in this system is dopamine. Dopamine neurons with

---

* That means pressing the lever more than once every two seconds – and keeping it up for a full hour!

their cell bodies in a region called the ventral tegmental area extend into the nucleus accumbens and the prefrontal cortex. These regions work together to drive us to seek rewards. When we experience something pleasant, the dopamine levels in our nucleus accumbens increase, so it is easy to assume that this is the experience of pleasure. However, careful experimentation has begun to suggest that this might not be the case. Instead, dopamine might be more important for other aspects of reward, rather than enjoyment.

Monkey see, monkey do(pamine)

In the 1980s and 1990s, Wolfram Schultz carried out a series of experiments which suggested dopamine wasn't rewarding in itself, but was important for reward learning. It is vital we can recognise a juicy reward when we experience one, but it is arguably more important to be able to learn what predicts that reward. For example, knowing ice cream is tasty is all very well, but if you can learn that a tinny rendition of 'Greensleeves' means an ice-cream van is around the corner, you are much more likely to get ice cream again in the future. This kind of learning is a type of classical conditioning, famously shown in Pavlov's experiment. Just as his dogs learnt that the sound of a bell predicted food, and began salivating, we can learn that on hearing the chimes of the ice-cream van we should approach it, money in hand, and order a cone.

Schultz was working with monkeys, and developed techniques that allowed him to record specifically from the monkeys' dopamine neurons while they were awake and behaving. He discovered that these neurons were active when the monkeys received a reward like fruit juice, much as you might expect from a neurotransmitter involved in pleasure or reward. But then he showed monkeys a range of patterns and trained them to expect juice whenever they responded to a particular one. And he saw something surprising. Once they had learnt this association, their dopamine neurons no longer responded to the juice itself, instead becoming active when they saw the correct pattern. Were the monkeys experiencing pleasure from simply seeing something that predicted a reward? Or was the dopamine actually coding something different: a prediction?

To try to work out what was going on Schultz tried a few more variations. He found that if they got juice when they weren't expecting it (i.e. to a pattern that had never before predicted the reward), the monkeys' dopamine neurons became active again. And if the monkeys were expecting a reward and none was given, their dopamine levels actually reduced.

Because of this experiment, and the many that followed it, Schultz believed he had discovered a different role for dopamine, not reward as such, but prediction error. In every moment of our day we are predicting what will happen next, based on our past experiences. We aren't experiencing the world as it really is, but based on those expectations. Dopamine's role is to let us know when something is better than we expected, because unexpectedly good outcomes are worth remembering, to ensure we repeat them in the future.

Take, for example, the first time you tried a new food, something you were a bit sceptical about. If it was actually delicious, your dopamine neurons would be highly active, because the food was much better than you predicted. This would make sure you learnt that your initial prediction was incorrect, so that next time you saw that food you would approach and eat it. The dopamine was telling your brain it had made an error. Once you have learnt how much you like the food, however, dopamine won't respond as much when you eat it as there is nothing new to learn. And the same could happen for drugs. Your brain experiences a rush of dopamine, causing you to learn about the drug very quickly.

This also explains my experiences with my books. When I was first asked to write one, this wasn't something I could have predicted, so the rush of dopamine was large. But after that, as the requests came in for the next book, and the one after that, there was no longer a prediction error. It was an expected outcome, and so less dopamine was released in my reward circuit.

Despite these findings, in the 1990s, the case for dopamine being the pleasure chemical seemed strong. A lot of it was based on the research of Roy Wise. While working at Concordia University, Canada, Wise had spent decades experimenting on rats, investigating hunger and motivation. In one study, he used drugs to block dopamine and watched their response. Time after time, hungry rats seemed to get bored and give up pressing a lever that rewarded them with tasty food, behaving in the same way as rats for whom the lever no longer provided any food. Interestingly, the rats on the drug behaved normally until that first taste of food, leading to the conclusion that these drugs, as Wise puts it in his 1978 paper, 'take the "goodness" out of normally rewarding food.'

Around the same time, Kent Berridge was busy becoming a leading expert on rat facial expressions, and using these to tell whether they liked or disliked something. Just like us, he found, the rats would lick their lips when they tasted something they liked, and make a gaping face for bitter tastes, which they disliked. To check it wasn't just a reflex, he taught the rats that a certain pleasant taste made them ill, and saw that they started responding as if that taste had become unpleasant.* Wise learned about Berridge's technique, and in the late 1980s, while Berridge was a professor at the University of Michigan, asked him to collaborate. The pair hoped to cement the idea that without dopamine, the rats wouldn't experience pleasure when given tasty food. As Berridge told me, 'We expected that if we supressed the dopamine system, these liking reactions would go down.' But as so often happens in science, the results were not as the pair expected. While the dopamine-deprived animals wouldn't choose to eat the food, if you placed it in their mouths, they still showed pleasure responses. 'A rat who wakes up in this state will not voluntarily eat, it won't voluntarily drink, it won't respond to any reward. And we could ask them,† do they still like the taste? And they did, they were absolutely normal. So then we could ask them, can they still learn new likes and dislikes for different tastes… and they could learn new things.'

---

* The same reason I still can't drink rum and Coke…

† Obviously, he doesn't mean he actually asked them – as far as I know scientists are yet to genetically engineer a talking rat! Berridge's techniques allow him to read the facial expression of his rats, and so understand whether or not they were experiencing pleasure when, for example, eating tasty food.

The finding that animals with no dopamine can still learn about rewards showed that, like so many others, Schultz's prediction error theory can't fully explain the workings of our complex brains. There were other problems for it too. For example, it isn't just learning that affects how much an animal values a reward; their internal states are also important. Very salty water is normally disgusting, but if a rat is severely salt-deprived, something that predicts the arrival of such water (often called a 'cue') suddenly becomes very attractive, and activates their dopamine systems, without them needing to relearn its association with the reward. And, perhaps most convincingly, if you boost dopamine in the brains of rats, they don't respond by learning more or faster. Instead, things simply become more tempting.

Hedonic hotspots

So prediction error couldn't be the whole story, but the dopamine pleasure theory didn't fit Berridge's results either. Stunned, and not quite believing his own findings, he put forward a new idea. Perhaps dopamine was responsible not for how much an animal *liked* something, but for how much they *wanted* it. To see if this was the case, he carried out more experiments. 'I didn't really believe the wanting versus liking then – we suggested it as a possible explanation. But then we went on to turn on the dopamine system, and we could turn on an intense want, make rats eat more, make them work more for food, and again the liking was unchanged.'

Berridge went on to find that even in rats that had had 99 per cent of their dopamine-making neurons destroyed, the pleasure response was intact. And these rats, whose condition was very similar to that seen in late-stage Parkinson's disease, could still learn new associations, such as tastes paired with sickness. Humans with Parkinson's also report still enjoying their food, even when their dopamine levels are severely impaired.

So if it wasn't dopamine in the reward system that caused feelings of pleasure, what was it? Working with Berridge, Morten Kringelbach has been looking into just this. Together, Berridge told me, the pair have found 'hedonic hotspots' in a rat's brain. 'In spots of the nucleus accumbens where dopamine is going, there is a little spot we call a hedonic hotspot where we can turn on intense liking.' By stimulating these areas, the researchers found they could boost an animal's enjoyment of sweet foods, causing the rat to lick its little lips even more in pleasure.

Unlike the 'wanting' system, these hedonic hotspots don't rely on dopamine. Instead, they use two different neurotransmitters, both our brain's natural versions of common drugs. One of these, enkephalin, is an opioid, similar to morphine. The other, anandamide, is like the active ingredient in cannabis, one of a class of compounds called endocannabinoids. When we do something enjoyable, these two chemicals are released in these hotspots, creating a feedback loop where each seems to boost the release of the other, and it is this that gives the sensation of liking something.

As with everything in the brain, these pleasure hotspots don't work alone. They are connected to a region at the very front of the brain that produces the conscious feeling of enjoyment, and also regulates these sensations. A small area within this 'orbitofrontal cortex', just behind the eyes, seems to be responsible for making an enjoyable taste become less pleasant – for example, when you have eaten too much of it.

So do drugs of abuse activate this pleasure system, and is that why people take them? Heroin and prescription drugs like morphine and oxycodone mimic our bodies' natural opioids, which we have just seen give us feelings of pleasure. They are also the body's painkillers, released to damp down pain when we experience something traumatic. But these drugs contain much higher doses than we would ever experience naturally, making the rush of pleasure they produce more intense than anything a user will have experienced before. This is what, initially, drives many to repeat the experience.

Maybe it's this simple effect. People become addicted because they enjoy the feeling, it's a simple reward mechanism. But that wouldn't explain why it's possible to become addicted to nicotine, which doesn't produce euphoria or even really pleasure. And studies have shown that people with addiction seem to want a drug that doesn't have any noticeable effects. In one experiment by R. J. Lamb and colleagues, five people addicted to heroin were placed in a room with a lever. If they pressed the lever enough times, they were given either a dose of morphine, or a placebo. While all the participants said they enjoyed the highest dose of morphine given, they reported no difference in how they felt when given a low dose – there were no positive sensations caused by the drug. But although they *reported* no difference, they did behave differently depending on the dose. Fascinatingly, when only the placebo was being given out, they would quickly stop pressing the lever. But they would continue pressing for the low dose of morphine, even though they couldn't consciously feel its effects. So there must be some other process going on.

Perhaps it is the enjoyment of the drug that drives use initially, but once a user has taken drugs a few times, they might also continue taking them to avoid the negative sensations associated with withdrawal. These ideas fit with some of our understanding of psychology. If something is pleasant, we learn to seek it out and do it again. If it is unpleasant, we try to avoid it. The 'come-down' from many drugs is seriously unpleasant, because our brains can rapidly become dependent on them. Might this be enough to explain addiction?

One of the classes of drugs that causes the biggest dependence reaction, most rapidly, is opioids. When someone takes heroin or a related drug, the amount of opioids in their system is much higher than would ever happen naturally. The brain isn't used to these huge doses of these chemicals, so it begins to fight back. The neurons that respond to opioids begin to become less sensitive, to counter their effects. This is known as tolerance, and means the brain can function more 'normally' when the drug is in the person's system, but it also has two knock-on effects. First, it means that larger and larger doses are needed to produce the same feelings of pleasure. Second, it means that when the drug hasn't been consumed, natural rewards (which cause a small amount of opioids to be released) are likely to have very little effect at all. With things like food and sex no longer producing the pleasure they once did, a user is even more likely to seek out the drug again, to recapture that feeling.

As well as activating these opioid receptors, heroin and related drugs have other effects. They cause the release of huge amounts of dopamine from the ventral tegmental area, and they suppress the release of noradrenaline in an area called the locus coeruleus. Noradrenaline is a hormone involved in the body's fight-or-flight reflex, speeding up our heart rate and breathing to prepare us to tackle a threat (see Chapter 4). It also functions as a neurotransmitter in the brain. Here it helps us feel alert and focuses our attention, but can also produce feelings of anxiety. Suppressing this chemical leads to some of the symptoms of opioid intoxication, such as drowsiness, relaxation and slowed breathing. Just like the opioid neurons, with repeated use, these noradrenaline neurons begin to fight back, producing more of the neurotransmitter, so that when a user takes the drug, the effects aren't so great – they have developed tolerance. But, of course, this means that when the drug leaves their system there is nothing to damp down the effects of the noradrenaline, and they experience withdrawal symptoms such as jitters, muscle cramps and diarrhoea.

Something similar occurs with other drugs, including caffeine (see Chapter 5) and alcohol. But the peak for withdrawal symptoms usually occurs only a few days after the person stops taking the drug, and lasts at most a couple of weeks. After this point, the neurons in their brain return to their usual set-point, releasing a normal amount of neurotransmitter. So if withdrawal-avoidance was the only reason people continued to take drugs, at this point they would be cured, and no one who successfully came through a detox programme would ever relapse. But this is far from the truth, as Berridge told me: 'The fact is that people come out of withdrawal, and yet the addiction doesn't seem to fall away. They are still very vulnerable, especially at moments of stress, which activates the dopamine system.' In fact, many people with addiction relapse years after their last fix. That's why they often refer to themselves as 'recovering', rather than 'recovered', for their entire lives. So addiction must be something more than just trying to avoid the nasty effects of withdrawal. There are also drugs that don't produce withdrawal effects at all, like amphetamines, but people can still become addicted to them. Clearly, addiction and drug use are far from simple processes, so it seems we need to dig deeper into the brain's reward and motivation circuits to understand what is happening.

Work it, baby

Back in the 1980s and 1990s, at the University of Connecticut, there was another researcher who wasn't convinced that dopamine was a pleasure chemical. In a set of experiments, John Salamone gave hungry rats a choice between bland 'rat chow' and tasty, sugar-rich food. Over and over again, as you might expect, they chose the sugary option. But then he made the decision a little more difficult. The rats could eat the chow without putting in any work, but to reach the sugar, they had to clamber over a barrier. Using this technique, Salamone could find out exactly how much the sugar was worth to the rats based on how much work they were willing to put in to get the better reward.

Salamone had first developed an interest in reward and motivation while working on his dissertation in the UK. At this time, the received wisdom was that dopamine was a pleasure chemical, released in the brain when an animal did something enjoyable, such as eating tasty food. But he had a sneaking suspicion that dopamine's role was more to do with having the energy and motivation to reach your goal, rather than enjoying it when you got there. As he explained to me, 'In chemistry there is this idea of "energy of activation" – in order to get a reaction to work you have to put energy into the system. And I sort of made this analogy that in order to get significant things we have to put energy in, we have to act. And I thought that that was close to what dopamine systems were doing.'

A few years later, and now a professor at the University of Connecticut, he decided to find out. As well as the experiment with sugary treats, Salamone tested different types of reward (such as a large pile of food versus a smaller one) and different types of work, like a certain number of lever presses. Then he began to change the amount of dopamine in the animals' nucleus accumbens. He found a robust pattern. If dopamine levels were increased, the animals would work harder for the preferred reward. If they were decreased, they would tend to go for the easy option.

This was only the case when there was work involved; dopamine-depleted animals would still choose the good reward if all other things were equal, and still seemed to enjoy the tasty food when they were given it. But the dopamine levels had changed something. In effect, it altered how motivated they were to go for the big pay-out, or how much they wanted it. As Salmone told me:

> The animal is still hungry, they are still food motivated. But they tend to be less likely to work for it. If you eliminate the work requirement, you don't change

> anything, they eat the same amount and they still have a preference. We've also shown that this is bidirectional – you can give drugs that increase dopamine transmission and increase selection of the high-effort option.

Animal studies are fascinating and can allow us to understand neuroscience at a basic level that we just couldn't achieve without them, but there is always one big concern. We aren't the same as rats. So how do we know that dopamine is as important for our motivation as it is for theirs? To try to prove his theory applied in humans as well, Salamone began collaborating with clinical psychologist and neuroscientist Michael Treadway.

Treadway devised a version of Salamone's task suitable for use in humans. Rather than food, they would earn money but, just like the rats, they would have to work for it. The participant's challenge was to press a button repeatedly, raising a bar on their computer screen. But they had a choice. They could either use the first finger of their dominant hand to make 30 presses, for a small reward, or the little finger of their non-dominant hand to make 100 presses,* for the larger reward. The amount of money they could earn could be varied, and there was another element: the subjects weren't guaranteed a win even if they completed the task, and the likelihood they would win could be changed. So, the researchers could compare, say, a 12 per cent chance of winning $1 for the easy task or a 12 per cent chance of winning $4 for the hard one with another option where there was an 88 per cent chance of winning the same amounts.

* This may not sound too bad, but it is surprisingly difficult – try it!

In their first experiment using this technique, Treadway and colleagues found that students who scored higher in anhedonia (a trait which means you enjoy things less, and is one of the most common complaints of people with depression) were less likely to work for the larger reward, particularly when there was a high level of uncertainty around winning the reward. Previous studies have linked anhedonia with low levels of dopamine, so this was a good initial indication that dopamine is important in this form of motivation in humans as well as in rats.

Along with mood symptoms, one of the most common complaints in people with depression is a lack of energy. They lose their desire to do activities that once brought them pleasure, feeling like they are just too tired, and that the reward of doing the enjoyable thing just isn't worth the effort it would take. So, the pair decided to try their experiment on people with depression. 'Depressed people also have a low-effort bias, similar to our animals. I started thinking about developing this as a form of animal model.' This led Salamone to his current work, using dopamine-deprived rats to mimic the motivational symptoms of depression. But, he tells me, it's not just depression that has this lack of energy as a symptom. It might not be what comes to mind when you think of schizophrenia, but what are known as 'negative symptoms', such as anhedonia and low energy, are a huge part of the condition for many people. And for many other conditions, this same symptom is given another name – fatigue.

In a clinical sense, fatigue isn't the same as the tiredness healthy people experience at the end of a hard day's work, or after going for a run. It is an all-encompassing lack of energy, a feeling that you just can't get going to do the things you want to do. And it's a feeling I know well. Ten years ago, I came down with a virus. Suddenly I was completely drained of energy, going from dancing six times a week to struggling with one flight of stairs. When this, and the host of flu-like symptoms that went with it, didn't get better in a couple of weeks, I went to the doctor. After a couple of months, I was diagnosed with post-viral fatigue syndrome. After six months, still no better, it was upgraded to chronic fatigue syndrome (also known as myalgic encephalomyelitis, or ME/CFS). Although I am a lot better now than I was in those early days, I still struggle with fatigue every day, so this area of research is of real personal interest.

Fatigue is also a symptom of other conditions like multiple sclerosis (MS), and a side effect of chemotherapy and treatments given for many conditions, including Huntington's disease. Huntington's disease is a genetic condition which causes brain cells to progressively break down and is, currently, uncurable. As the disease progresses, people often suffer from jerky, uncontrollable movements, and a drug called tetrabenazine can be given to help reduce these symptoms. How exactly it helps isn't fully understood, but the drug works by reducing the amount of a group of neurotransmitters called monoamines that are available to be released into the synapses of neurons. This includes dopamine. One of the major side effects of this drug, along with depression, is fatigue, so Salamone began to wonder if it could be used to find out more about the mechanisms of this symptom, and maybe even develop treatments. 'We found you get a very robust shift in effort-related choice with tetrabenazine. And then we started looking at things like what kinds of drugs improved the performance in a tetrabenazine-treated animal. And one of the things we looked at was SSRIs, because they are antidepressants, so they should work. But they didn't reverse its effects.'

SSRIs, the most common form of antidepressant, boost the amount of serotonin available in the synapses. But as they have no effect on dopamine, it's perhaps not surprising they didn't help with motivation. Salamone's findings exactly mirror the experiences of many humans with depression who are given these drugs. It is common to find that while their mood does improve, fatigue symptoms don't. Another drug, called Bupropion, which affects both dopamine and noradrenaline, however, did change the rats' behaviour.

Another way Salamone found that he could induce this fatigue state in the rats was to give them proinflammatory cytokines. Cytokines are small proteins which carry messages between cells in the body, and in humans proinflammatory cytokines are produced in response to illness or injury. In the short term, they are beneficial, helping with the healing process, but they do have another effect, which Salamone has explored by giving healthy people these molecules. 'In humans they produce something like that fatigue or lack of energy even more readily than mood dysfunction.' One of their effects is to produce the general malaise that we feel when we have an infection: the brain fog, tiredness and lack of desire to do anything except lie on the sofa and watch daytime TV. Salamone believes this may actually be an adaptive response: 'You save your energy resources, but you also disengage from social interactions and so you are less likely to spread the infectious agent. So that might underly the evolutionary value of having this link between peripheral inflammation response and why you go into this pattern of behaviour.'

But, in some cases, this system can go wrong, triggering long-term inflammation, and this has been linked with illnesses from ME/CFS to depression. We don't know exactly how cytokines affect the brain, but there is some evidence they reduce the synthesis of dopamine, again fitting with Salamone's findings. Salamone's next hope is that his work might lead to the development of drugs to treat some of these conditions. He has already found that the effects of these cytokines in rats could be reversed by giving them the right drugs, to boost their dopamine levels or bypass the dopamine system entirely and exert an effect on the neurons in the nucleus accumbens via a different neurotransmitter. If this finding translates to humans, it could improve the lives of people with depression, schizophrenia and the myriad of other medical conditions which have fatigue as a component. But, of course, this is the human brain, so things are rarely that simple. Boosting dopamine in the brain, it turns out, can also have its downsides.

Mo' dopamine mo' problems?

Currently, dopamine-boosting drugs are most commonly prescribed to people with Parkinson's disease, to try to relieve their symptoms. But, as with most drugs, they can have side effects. Some of the most surprising of these are grouped together as 'impulse control disorders'. Some people suddenly start gambling, or binge eating, while others become hypersexual or develop a compulsive shopping habit. What groups all of these issues together is that they are driven by intense cravings, which the person struggles to control. This fits nicely with our understanding of dopamine so far. By boosting levels in the brain, these people have become more reward-driven, more motivated to seek out their high of choice, whether that is a flutter on the gee-gees or a new pair of shoes. In some cases, they can even become addicted to their medication, taking more than is needed to control their symptoms. These cases are, thankfully, rare, but they do raise concerns about using dopamine-boosting treatments for other conditions. And they do add support to the idea that dopamine might motivate us to seek out rewards, however damaging they might be in the long term.

When it comes to drug use, this finding, that increasing dopamine in the brain increases reward motivation, is an important piece of the puzzle. While drugs that can be addictive contain a wide variety of chemicals, and act on the brain in a myriad of ways, many of them have one thing in common: they cause the ventral tegmental area to release large amounts of dopamine.

Back in Berridge's lab, the team have been working on the effects of this for many years. As we have seen, they found that the dopamine system increases how much animals want a 'reward'. But, he explained to me, this is a very special kind of seeking. 'There is a particular psychological signature feature to this kind of wanting. It's not like all wanting – it's really linked to cues. When something catches your eye and you can't look away, you want that – that's this dopamine system.' We have all experienced it. Going about your day, not feeling particularly hungry when you walk past the entrance to a bakery… BAM! The smell of freshly baked bread hits your nostrils and you are instantly starving, and unable to resist one of their delicious fresh rolls or pastries. This is dopamine's power. It produces instant, strong desires, based on your environment. As Berridge explains, 'Your desire for tasty food is very different from your desire for world peace.'

When you take a drug, several things happen in your brain. As we have seen, we can develop tolerance to the effects of a drug, which can lead to dependence. But surprisingly, in some situations the dopamine system doesn't seem to develop tolerance and instead we see what is called *sensitisation*. As we saw in Chapter 2, sensitisation can make sea hares respond more to something that before wouldn't have elicited much of a response at all. Kandel's work went on to form the basis of our understanding of long-term memory, but it is also remarkably relevant to drug addiction.

When an animal is given a drug like amphetamine or cocaine, they become restless, and measuring their movements can give an idea of how much the drug has affected them. This is directly related to dopamine release. Just as too little dopamine can produce the rigidity and difficulty in making voluntary movements seen in Parkinson's, too much can produce the opposite effect. Studies on rats and humans have shown that in some cases more dopamine is released when they are given a drug for the second, third or even fifth time. Shockingly, this still happens even when a user has been drug-free for over a year.

In one study, rats tested after 30 days of withdrawal from cocaine showed this increase in movement when given the drug, suggesting a sensitisation of their dopamine systems. This could be an important finding in addiction research and, as Berridge explained, it may even explain why some people are more likely to relapse after detoxing from a drug. 'Some individuals are really susceptible to this sensitisation. The dopamine system isn't always hyperactive, but it's hyper-reactive, in a virtually permanent sense, that can last for years even if you don't take drugs. So it's reactive to the drug and it's reactive to drug-paired cues.' A sensitised reward system is like a precarious pile of washing-up, primed to come crashing down at the slightest provocation.* With a dopamine system set up like that, the tiniest reminder of a previously abused drug can trigger dramatic cravings.

While this work is fascinating, it is important to see whether it can be reproduced in our own species before we make claims about sensitisation being important for addiction. And that is exactly what Marco Leyton is working on at McGill University, Canada.

Interestingly, many of the initial attempts to reproduce these effects in humans weren't successful. In fact people who had used a drug for the longest time seemed to have a reduced amount of dopamine release when they were given the drug in a lab (which would fit with ideas about tolerance, but not sensitisation). Looking at a number of these studies, Leyton realised that the dose of drug given changed everything. When they took a large

* Or, in our house, like a badly stacked Tupperware cupboard, where pulling out one pot causes an avalanche of plastic.

enough amount, participants showed exactly the same response as the rats. One benefit of studying humans over animals is that you can ask them questions about how they feel, and he also noticed that the questions asked were important. When studying amphetamine, for example, the sensation that seemed to be most reliably sensitised was a feeling of energy.

Leyton set about carrying out his own experiment, working with Isabelle Boileau, Head of the Addiction Imaging Research Group at the Centre for Addiction and Mental Health, Toronto. With their team, the researchers took a group of students who had never taken amphetamine (or any stimulant drugs), gave them a dose of amphetamine and used brain imaging to see how much dopamine was released in the striatum (an area of the brain that includes the nucleus accumbens). They then gave them two more doses over the next few days, before sending the participants away. When they returned for a fourth dose two weeks later, they found significantly more dopamine was released than in the initial study, and the students reported feeling more alert and energetic. Amazingly, when they tested them for a fifth time a year later, there was an even bigger effect.

So what does this tell us about substance users? To make sure that these effects worked on people who have been using drugs for a long time, rather than those who had only taken drugs in a lab setting, Leyton studied a group of cocaine users and a group of amphetamine users. He found, as expected, that the more cocaine the users had taken throughout their lives, the more they showed this sensitisation effect. But the amphetamine users actually showed the opposite effect, with the most regular users having the smallest dopamine release.

Leyton realised he had discovered something else. It appears that sensitisation needs drug cues in order to work. His cocaine group had been given a mirror and a white powder and were allowed to prepare and take the drug as they normally would, so their brain experienced a lot of the same cues it would normally before they took a drug. The amphetamine group, however, just got a nondescript pill.

This led him to put forward a proposal that dopamine is dysregulated in drug users in two ways. If there is nothing drug-associated around, too little dopamine is released, so they can't sustain interest in other rewards, like food or social interactions. But whenever something in the environment reminds them of the drug, their dopamine systems go haywire, so people with an addiction end up with an overwhelming desire to take the drug again, and will work hard to seek it out. This desire continues even when they stop liking the drug, and can overpower all other wants and needs in life.

Back in Michigan, Berridge's current work is looking at how one thing becomes our current greatest desire, over and above another. For most of us, this might be the switch from hunger to thirst, but for people with an addiction it is this process of the drug taking priority over everything else. Berridge says:

> There is one big question in addiction neuroscience, which is: why is it that we want one thing so much more than others, what is causing the focus of the wanting? You or I want food when we are hungry but not when we are

> thirsty, and we switch back and forth throughout the day. But an addict may have it more fixedly stuck on the addicted drug.

Using a technique called optogenetics, Shelley Warlow and colleagues in Berridge's lab are able to activate certain neurons in a rat's brain painlessly, using light. This allows them to switch on the dopamine system and see how it affects the animals' behaviour. They have found that they can use this to change an animal's wants, making them choose a specific sugar treat or cocaine-dispensary over another identical one. They have even found that this brain stimulation is enough to make rats repeatedly approach and investigate a bar that gives them a painful electric shock when touched. Normal rats will learn after one or two exploratory touches to stay away, and will even try to bury the offending item. But with their dopamine systems switched on, Berridge's rats will keep coming back for more. Although they still find the shock painful, in his words: 'The rat becomes fascinated by the rod. It just hovers over it, sniffs it eagerly, touches it with its paw. It gets the shock and darts back, but then comes back and will be even more excited and interested… it just cannot resist the attractiveness of this rod as a cue… they just can't seem to stay away.'

Welcome… to Rat Park

But it's not just sensitisation to drug cues that makes it difficult for people who have been addicted to a drug to stay clean. Stress also plays a role. Studies have shown that in sensitised people, stress can trigger a release of dopamine just like a drug cue can, which may be why emotional life events, whether positive or negative, can often be triggers for relapse. Your genes are also important. Users with a family history of drug abuse have been found to release less dopamine in their striatum when there are no drugs around, but more when they are exposed to drugs. People who score higher in 'novelty-seeking' personality traits are also more at risk of sensitisation.

Despite these findings, it's not just brain chemistry that controls whether someone becomes addicted, and whether they can kick the habit. The environment plays a role too. Early studies, which showed rats repeatedly self-administering drugs, housed the animals in small individual cages, with nothing much to do except take the drugs. Is it any wonder this intelligent, social species resorted to getting high?

In a series of experiments known as 'rat park', Bruce Alexander and colleagues at Simon Fraser University, Canada, found that if the rats were instead kept in groups, in large cages with running wheels, platforms for climbing on and plenty of places to hide, they consumed a lot less morphine when it was offered.

Evidence that this also applies to humans comes from history. All over the world, when Europeans colonised new lands, the indigenous people were treated appallingly. They were forced out of their homes and moved into 'reserves', forced to forgo their own culture for that of the colonisers, and even, in some cases, had their children taken from them. In many of these displaced groups, addiction rapidly took hold. Even today, Indigenous populations in Australia are still more likely to have substance abuse problems. Statistics from

2018 show that while they make up just 2.8 per cent of the Australian population, Indigenous Australians made up 16.1 per cent of clients of alcohol and other drug-treatment services.

We mustn't over-hype the findings of the rat park experiments, however. They weren't perfect, as no experiments are, and there are ways bias might have been introduced. Follow-ups and replication attempts have had mixed results, with some finding no difference or even more drug-taking in the rat park residents. While this doesn't mean we should throw out the original results, it does suggest they aren't as robust as some writers might like us to think. It is clear that the environment does play an important role in addiction, but it isn't the only factor. Humans are incredibly complex, and our behaviours arise from a number of different factors, including genetics, brain structure and chemistry, and our current and past environment, all of which can also impact on the others.

The big question now is whether our developing understanding of the brain basis of addiction could help improve treatments for this devastating disorder. Sadly, but perhaps in hindsight not surprisingly, the (now outdated) concept of dopamine as a reward chemical didn't lead to any successful treatments. And so far, neither has the new concept of dopamine driving drug cravings. One reason for this may be that it is an oversimplification. Most studies in this area are done using stimulant drugs, which do cause dopamine release in the striatum. But in a recent review paper, David Nutt, Professor of Neuropsychopharmacology at Imperial College London and ex-government advisor on drugs, argues that other drugs, like heroin, don't cause dopamine release. And alcohol and nicotine, the two most common addictive substances, may have different effects again. Studies are mixed for all of these drugs, but Nutt sums up the issue nicely in his paper, writing:

> Addiction is a complex mixture of behaviours and attitudes that vary from drug to drug and from user to user, and it is unlikely that a single neurotransmitter could explain every aspect of addiction... Unifying theories, although intrinsically appealing, should be subject to careful scrutiny just like other theories – and perhaps even more so because they can lead the field into directions that ultimately prove to be unfruitful.

If the factors involved in addiction are so complex, it isn't really surprising that our day-to-day drives and desires can seem mysterious. Motivation can be influenced by a huge number of factors, such as mood and hunger levels (both of which we will cover later in the book). But understanding the basics of the reward system can help us to exert control over our more basic instincts. And this is where the frontal lobe comes in.

There is a problem in psychology and neuroscience, and it is largely one of language. When I talk about 'us' exerting control over 'our' more basic instincts, in both cases I am referring to our brains. We are our brains, and they are us. But the area where our consciousness seems to have the most control is the frontal lobe, behind our foreheads. This is the area that is over-developed in humans compared to other animals, and it is where our reason and what is known as 'executive function' lies.

This includes our ability to control our attention, plan for the future and, most importantly in this context, inhibit inappropriate actions. It is also the last area of the brain to develop, not finishing maturing until our late twenties. It is this region that controls impulses, damps down emotions and channels our motivation in the right direction. In people with an addiction, it can become impaired, making it harder to resist the lure of the drug. But, in some cases, people can recover.

People recovering from addictions are fighting against a myriad of changes to their brains, affecting the reward and motivation circuits, frontal lobes and other areas. These changes mean the odds are stacked against them. But despite that, in some cases, with the right support and in the right environment, people can change their brains.

If this is possible for someone with something as serious as a substance addiction, just think about what that means for the rest of us. The more we know about our in-built drives, to keep moving forward and aiming for that next reward, the more we can practise exerting control over them. We can use this motivation when it is beneficial for us, like striving for a promotion at work, and control it when it isn't. But while this is possible in principle at least, it is far from easy. And there is one major factor that can throw a spanner in the works of our brain's best-laid plans. Our emotions.

CHAPTER FOUR

Mood Swings and Scary Things

A few years ago, I was lucky enough to travel to the Galapagos. Not only is this collection of islands beautiful and (relatively) unspoilt, it is also home to wildlife found nowhere else on Earth. While travelling here in 1835, a young Charles Darwin collected species of finch from different islands. He noticed that their beaks were different shapes, but it wasn't until ten years later that he realised the importance of his finding. Each bird, descended from one common ancestor, had evolved and specialised for the kind of food that was found on their particular home island. This helped inspire his theory of evolution by natural selection, and as a huge geek, I was incredibly excited to follow in his footsteps, and see some of the animals he wrote about for myself.

The trip didn't disappoint. Each day, our ship stopped at a different island,* and small groups of us climbed aboard a dinghy and were taken the short distance to shore. Some of the islands had perfect white sandy beaches, while others were craggy, bare volcanic rock, or covered in greenish red scrub and tall cacti. My favourites of the animals were the marine iguanas, so unbothered by their human admirers that we had to

* Most of the islands are uninhabited by humans, so staying on a cruise ship is the only way to see the majority of them.

climb over them where they lay basking on the paths. Once or twice I nearly stepped on one! I also have a huge soft spot for blue-footed boobies – funny-looking birds with bright blue feet, which have the most wonderful courtship ritual.*

We also spent some time snorkelling amongst brightly coloured tropical fish, and incredibly playful sea lions, who would blow bubbles and then chase them to the surface. And it was during one of these snorkelling trips that I learnt something rather unexpected about myself. We were swimming in a shallow bay, surrounded by mangrove forests. The water was murkier here than in other places, so visibility was only a couple of metres, but I was enjoying watching the fish and other creatures below me. Meanwhile I was keeping an eye out for something more exciting: a shark. Mangrove swamps often form nurseries for baby sharks,† because the roots keep out larger animals that might eat them, while fish and other small prey animals are plentiful. I was excited at the possibility of seeing a shark up close, but while I wasn't exactly planning on swimming up and hugging it, I wasn't prepared for the fear that would flood my body when the shape began to emerge from the gloom.

---

* My husband actually used part of their ritual when he proposed to me. A male blue-footed booby presents the female with a particularly nice stick he has found, something that is thought to be an evolutionary hangover from when these birds built nests. It is now purely symbolic, as they lay their eggs in the sand, but remains (I like to think) as a promise that the male will look after his female. Although Jamie couldn't find a stick, he used a mangrove pod instead – he would never have cut it as a booby!

† Do do do do do… sorry, couldn't resist!

I have never been afraid of sharks, mainly because I know the stats. In the US, for example, you are 30 times more likely to be killed by lightning than by a shark bite. And these were babies, for goodness sake, hardly longer than my arm. Even if one did decide to have a nibble, it wouldn't do me much harm. The rational part of my brain knew all this, but it wasn't enough to quell the primal instinct that felt like it was coming from deep inside me. It was, however, enough to stop me acting on it. I was able to damp down the fear enough to enjoy watching the baby sharks. By this point there were three of them, or maybe four, it was hard to tell as they kept appearing and disappearing in the murky water. At one point, I found myself in the midst of a shoal of tiny silvery fish, dapping and darting to avoid the sharks. Again, my brain and body screamed at me to get out of there, while the rational part of me tried to reassure me it would have to be a very stupid shark that confused me for one of those fish, which were just a couple of centimetres long.

But my biggest fear in that moment was probably the least rational. There was a tiny part inside me that kept expecting something bigger to loom into sight: one of the baby shark's parents. I knew rationally that this was almost impossible. Shark parents don't stay with their young, and they don't tend to come so close to shore. But I just couldn't shake the physical sensation of fear, and thinking about it still sends a shiver down my spine. This experience led me to wonder why this happens. How can our emotions become so strong that they can overwhelm the logical, rational parts of our brains? To find out, we are going to need to start at the beginning, with what emotions are, and why we have them.

Feeling the fear

While we can't know for sure, it seems likely emotions evolved to drive us towards rewards and away from danger. The clearest example of this is fear, the emotion I felt when seeing the sharks. When faced with a threat, our ancestors had two main options: they could run away from it, or fight it. The fear response has evolved to help us do one of these two things. When we see something that might be dangerous, signals travel from our eyes directly to our amygdala, part of the brain involved in emotional processing. This starts the bodily sensations of fear known as the fight or flight response.*

At the same time, the information also travels the longer route from our eyes to our cortex where the more rational parts of our brain can analyse it. It is only when it reaches this area that we become conscious of the threat. Have you ever felt a chill of fear from seeing a coiled shape in long grass, only to realise almost immediately it is just a garden hose? That feeling was the fight or flight response being triggered by the amygdala, then immediately quashed by the cortex. Once the information reaches this part of the brain, we can analyse whether the threat is genuine, or whether it is just a harmless piece of hose. But that feeling of panic hits first, just in case.

To initiate the fight or flight response, the amygdala activates the hypothalamus. This sends signals down your autonomic nerves, which control processes like digestion and breathing. The 'sympathetic nervous system' activates the fight or flight response, causing the adrenal glands to

---

* Or sometimes fight, flight or freeze, as some animals stop moving when faced with a threat.

release two related hormones, adrenaline and noradrenaline, into the bloodstream.* These travel throughout the body, causing a host of changes: your heart rate accelerates, breathing quickens, and blood flow to muscles increases to get as much oxygen to them as possible, in case you need to run. Extra oxygen travels to the brain, making you feel alert, and your senses are heightened. Blood sugar spikes as it is released from the liver to feed the muscles.

A few seconds later, the hypothalamus releases a chemical messenger, which through a chain reaction triggers the release of, amongst others, a hormone called cortisol (which we met in Chapter 2) from the adrenal glands, which sit above the kidneys. Often called the stress hormone, cortisol keeps the body on 'high alert', increasing blood pressure and blood glucose levels. It is cortisol that allows you to deal with stressful situations that last longer than a few minutes.

Meanwhile, any processes that aren't needed, and might waste precious energy, are switched off. Digestion slows down or stops, tear and saliva production reduce (producing the familiar sensation of a dry mouth) and precious blood is directed away from the skin, making you look pale. The immune system is supressed, and, in extreme cases, your bladder muscles relax. Releasing any urine you are carrying to reduce your weight might just give you that boost in running speed that makes the difference between life and death!† These two pathways, one rapid but short lasting, travelling through our nerves,

---

* These two chemicals often work together, but adrenaline is mostly found in the body, while noradrenaline is mostly found in the brain.

† Or, at least, that's what your body thinks.

and the other slower but longer lasting, carried by chemicals in our blood, mean the response can last as long as it needs to get us out of danger.

But this can be a problem. Nowadays, most of us rarely face the kinds of immediate threats our ancestors endured. Instead, our fight or flight reflex might be triggered by a big presentation at work, or money worries.* These stress-causing factors tend to be long lasting, and not something we can run away from, so levels of adrenaline, noradrenaline and cortisol can remain elevated much longer than they should. This is known as chronic stress, and it has huge impacts on health, from contributing to obesity to raising the risk of heart attacks and strokes.

As I write this book, many of us are experiencing chronic stress at a level we have never before endured, thanks to the Covid-19 pandemic. We are afraid of falling ill ourselves, and fear for our loved ones, particularly those who are more vulnerable. This is a real threat, so it is not surprising our stress responses have been triggered. But it is exactly the kind of threat our fight or flight mechanism is powerless against. Running away from the virus isn't going to help† and neither is physically attacking it. In fact, increased levels of the stress hormones can reduce the effectiveness of our immune systems, meaning worrying about the virus might actually make you more susceptible to it.

---

* Neither of which are cases where weeing yourself is at all beneficial!

† Unless, I guess, you are using your extra energy to dodge people in the park and ensure you stay two metres away from everyone at all times!

That is, of course, all very well to say, but shutting down the stress response is far from easy. It has evolved for a reason, and while there are ways to reduce its responsiveness, like using breathing techniques, exercise or focusing on what we are grateful for, we must also be patient with ourselves and realise it isn't surprising if we are feeling anxious at this weird point in history.

This explanation for why we have a fear response makes a lot of sense. If one ancient human always ran (or, I guess, swam?) away from sharks while the other headed towards them to try to boop their snoots, you can guess which one would be more likely to survive to become my ancestor. So the behaviour (running away) we produce in response to danger evolved to protect us.

But it doesn't tell us why we actually *feel* fear. It may seem to us as if it is our feelings that drive our behaviour, but it would, theoretically at least, be possible to have an animal (or robot) that produces bodily changes and behaviours in response to a threat, without the feeling we associate with fear. This is where we run into another problem of language. In general life, we use the word 'emotion' to mean both the bodily sensations and the feeling itself, but these aren't one and the same. In fact, it's perfectly possible to have one without the other – we all know someone who won't admit to themselves or others that they are angry or upset, while their behaviour shows they clearly are. To make it simpler, researcher and author of *Descartes' Error* Antonio Damasio suggests using the word 'emotion' to refer to the specific pattern of behaviour, and 'feeling' to refer to the mental state that usually accompanies it. I have tried to stick with that distinction throughout this chapter.

It might seem obvious to us that the feeling, the mental state, comes first – our heart starts pounding and we run away *because* we feel afraid. But our intuitions can be wrong, and this is where science can step in to help, by putting these beliefs to the test. In the mid-1880s, American philosopher and psychologist William James was interested in this issue and argued that we actually had the process backwards. In his theory, we feel afraid *because* we run. So, I saw the shark, my body reacted with the fight or flight reflex, and I interpreted those sensations as fear. A similar idea was suggested by physiologist Carl Lange at the same time, so it has become known as the James–Lange theory of emotion.

But there are problems with this theory. If the feeling of an emotion is entirely created by your body's physical reactions, there should be a distinct bodily sensation for each different feeling. If this were true, we would be able to tell exactly what someone is feeling just by measuring things like their heart rate, skin temperature or perspiration. But we can't; in fact, many emotions produce the same (or very similar) physical changes. And we should be able to make someone feel afraid, simply by giving them adrenaline. But, as we will see, that doesn't work either. Whether or not the original theory was more nuanced than it has become, it seems that the James–Lange theory isn't enough to explain the complexity of human emotions.

Scared stiff

In 1962, researchers Stanley Schachter and Jerome Singer put forward a variation on the theory. They thought that it was a *combination* of the bodily sensations and the environment that produced a feeling. To test this theory, the pair gave half their volunteers an injection of adrenaline, and half a placebo, telling them all that it was a vitamin they were testing to see if it affected eyesight. Participants were then placed in a room with an actor who was either acting in an amusing and entertaining way, or being very annoying. In both situations, the participants who had received adrenaline reported more emotional reactions, perhaps because they were ascribing the racing heart and shaky hands caused by the adrenaline to emotion.

The pair also found that if they told participants these symptoms would be a side effect of the injection, the influence on their feelings went away. They argued that this supported their idea that participants detected the bodily changes brought on by the adrenaline, but interpreted them based on their surroundings, and this combination was what created a feeling. If they knew that their heart rate was simply down to the injection they had been given, however, they could discount it.

Other studies have tested this idea in more 'natural' settings, to see whether bodily sensations associated with one emotion could be misinterpreted as linked to another. In one famous study conducted in 1974, Donald Dutton and Arthur Aron positioned an attractive female researcher at the end of a footbridge. They asked her to stop young men (aged around 18–35) walking on their own, who had just crossed the bridge. She asked them to fill out a questionnaire, and then write a short story based on a photograph. The researcher would then offer the men her phone number so they could call to find out more about the study later.

Unknown to both the men and the researcher, Dutton and Aron weren't actually interested in the effects of scenery on creativity, as they said. What they wanted to know was how many of the men would call. They found that the number calling depended on the type of bridge they crossed. When the researcher was positioned by a high, precarious suspension bridge,* which moved threateningly in the wind, a higher proportion of the men accepted her number and called her than when she was by a low, stable bridge. The stories of the men who crossed the scary bridge also contained more romantic and sexual content. To confirm their findings, they repeated the experiment with a male researcher (and still using male participants – I guess either not considering or not worrying about whether some were gay), and found the relationship disappeared.

This supported Dutton and Aron's idea. A range of emotions can be linked to the bodily sensation known as arousal. While we normally think of 'arousal' as sexual, in psychology it just means active, or 'ramped-up'. Think pounding heart, sweaty palms and quickening breath. Yes, this can be caused by sexual attraction, but it could also be caused by fear or anger. So, to work out what is causing the sensation, you look around and make an assumption. In this case, the men on the shaky bridge were misattributing the bodily sensations of fear as attraction, believing it was the woman who was causing them to feel that way, rather than the situation.

---

* For the pedants and engineers reading, including my husband, it is technically a simple suspension bridge. The actual one used in the experiments crosses the Capilano River in Canada and looks, from the photos, terrifying.

They then took the experiment into the lab. Male volunteers were told that they would be working with another participant, who happened to be an attractive woman. The experimenters flipped a coin, they were told, to determine which of them would get a horrible, intense electric shock, and which would receive a minor shock, which would barely be noticeable. While they were 'setting up', the men were asked to fill out a questionnaire, rating their anxiety and how attracted they felt to their female partner. As expected, the men who were told they would get the painful shock were more anxious, and more attracted to the woman than the men getting the smaller shock.

This makes a lot of sense. Emotions contain a bodily response, which tells you the intensity of the emotion, via sensory feedback, and a cognitive element, which identifies the type of feeling based on what is going on around you. There are now variants on this theory, but the majority of them use a term coined by Magda Arnold in the 1960s: appraisal. While I was swimming, I felt my heart begin to pound and attributed that feeling to the looming shark shadows, but if the situation had been slightly different, I could have thought it was my attractive snorkelling partner having the effect on me. This might be why scary movies are so popular on a date. Not only might your terrified crush grab your hand in fright, he might also think it is your charm and beauty causing him to feel nervous and flushed, rather than the flesh-eating zombies on screen.

Scientists still disagree about exactly how appraisal works in terms of which bodily cues are taken into account and whether these happen before, after or in parallel with the feeling. But new research techniques are helping us make sense of emotions. Scientists have found, for example, that people who are more aware of their own bodies (tested by asking them to sense their heartbeat without feeling for their pulse) experience more intense emotions. Brain-scanning studies have also shown the importance of the insula, a brain region involved in detecting and processing messages from the body. Although it is possible to feel emotions without it, its activation during emotional experiences in healthy people adds weight to the idea that our emotions are 'embodied', rooted in our physicality.*

This is where things start getting a bit philosophical. Because unless you believe the 'mind' is something separate from the brain or body (a theory known as cartesian dualism, after René Descartes), of course emotions must have their root in physiology! I certainly don't think the mind is some separate entity, but it is almost impossible to talk about emotions and consciousness without implying that it is. When I write that 'I' feel an emotion when certain areas of 'my' brain are active, what exactly is the 'me' that is doing the feeling? This is, of course, a huge topic, and one that is the subject of many books, so there is no way I could do it justice in a few lines here, but the puzzle of consciousness is so much a part of our understanding of emotions that I can't leave it out entirely.

---

* This is also something we have strong intuitions about, which we can see reflected in language. Phrases like 'butterflies in the stomach' or 'heart-broken' show how emotions seem to stem from the body.

Antonio Damasio argues that animals with simpler brains than ours could have similar 'emotional programs'. They may react in a certain way when threatened, or when they receive delicious food. But that doesn't mean they have feelings like ours. To really experience a feeling in the way we do, an animal needs to have consciousness and memory, and while some animals may well have an element of consciousness (this is another big topic), simpler ones probably do not.*

Does this mean animals don't have feelings, or moods? My own experience as a pet owner makes it hard for me to believe that. When I was young, we had a tortoiseshell cat called Emmy, who I adored. And she seemed to care about me too, even letting me touch her one beige paw, which, for some reason, would have triggered a hiss-and-swipe reaction from anyone else! She wasn't a particularly moody cat, but there was one thing which, like clockwork, would drive her into a huff: us going away. If we left her in the care of a neighbour for a few days while we were on holiday, our return would always go the same way. Initially, she would seem happy to see us, rubbing around our legs and purring. But within a few minutes, that would change. She seemed to be satisfied we were home for good, and it was time to teach us a lesson. For the next day or so she would completely ignore us (except, of course, when we put down food), giving me the furry cold shoulder if I tried to stroke her or encourage her to

---

*This also brings up the question of whether we could ever develop technology with such advanced artificial intelligence that it became conscious, and how we would know if it had. But that, again, is a topic for a whole other book!

play. But it wouldn't last long, and soon she would be back to her sweet, lovable self.

I may never know what Emmy was feeling at these times, if anything at all, but her reactions do seem to indicate emotional responses, and perhaps even something similar to the feelings and moods we humans experience when loved ones don't behave the way we hoped. There is also more scientific evidence that certain animals behave in a way that we would associate with feelings. Elephants appear to grieve and comfort each other after the death of a member of their group. Rats make noises that sound very much like laughter when tickled, and will encourage the experimenter's hand to continue if they stop. And macaque monkeys will resist pulling a chain to get food if doing so gives another monkey an electric shock. But despite this, it is impossible to really know whether they are feeling in the same way we do.

Moods, muscles and mussels

So far, we have uncovered, to some extent, what emotions are and what is going on in the body to cause them. But next we need to look at the brain. Brain research into emotions has focused on two main areas. The first is the limbic system, which includes the amygdala, and responds to emotional stimuli (particularly anything you are afraid of), and the hippocampus, which is involved in memory storage and processing, including emotional memories. The second is the prefrontal cortex, behind your forehead.

The amygdala and limbic system are reactive, and quick to respond to anything emotional, but they are (usually) kept in check by our more rational prefrontal cortex.

Most of the time, the two areas balance each other out, and your emotions stay relatively level. If you are faced with a threat, your prefrontal cortex allows the amygdala to take over and continue initiating the fear response, but only if the threat is a real danger. So, it is the prefrontal cortex that decides whether the problem is something you can handle, or something uncontrollable. It is this uncontrollable, chronic stress that causes problems.

Moods are a bit harder to pin down. Mood and emotion are words we use a lot, often interchangeably, but they are different. Emotions are short-lived, and usually triggered by something in the environment. You might, for example, feel happy because you got an invitation from a friend or angry because of someone mansplaining on Twitter. Moods are different. They are longer lasting, and while your surroundings can and do influence them, the effect is less direct. And moods are much harder to study. While it is pretty easy to make someone feel briefly happy or sad in a lab, by showing them pictures or asking them to listen to music, affecting someone's mood is harder. So a lot of the research into moods focuses on people with mood disorders like depression.

In depression, the same system seems to be important. Damage to the prefrontal cortex has all sorts of effects, including problems controlling moods, and various parts are found to be over- or underactive in mood disorders like depression. It seems that if the two systems get out of whack, emotions can start to overwhelm you, which can lead to low mood and, potentially, depression. This can happen because of an overactive limbic system, or an underactive prefrontal cortex. Either will tip the careful balance and cause prolonged release of stress hormones.[*] These can then, in turn, begin to damage prefrontal and limbic regions, leading to a feedback loop, which can be difficult to recover from.

But this still leaves us with some big questions. Why do these areas become unbalanced? And how are our brain chemicals involved? To find out, we need to dig a bit deeper, starting with those chemicals. And there is one more synonymous with depression than any other: serotonin. So that seems like a good place to start.

The discovery of serotonin began in the late nineteenth and early twentieth centuries, when scientists discovered an interesting effect of blood serum. This substance is what is left over when blood clots, and it was found to make blood vessels narrow. This wasn't hugely surprising. It was already known blood contained adrenaline, and that adrenaline could have this effect, known as vasoconstriction. What was surprising, however, was that the serum also made an animal's intestines constrict, whereas adrenaline alone had the opposite effect. This meant it couldn't be adrenaline in the serum that was causing vasoconstriction. It must be another substance that acted like adrenaline in most cases, but had a different effect on the intestines. They had discovered an 'adrenaline mimicking substance' in the serum.

---

* Interestingly, even something like blood sugar levels may be enough to tip the balance. Although evidence in healthy humans is limited, there are some suggestions that fluctuations in blood sugar cause the release of stress hormones and, over the long term, affect mood. Whether this can explain the short term experience of 'hanger', and excuse those of us who get irritable when we haven't eaten, isn't yet clear!

The next big step forward was made by Italian pharmacologist and physiologist Vittorio Erspamer in the 1930s. He isolated a substance from the digestive system of a rabbit which caused contraction of the gut, muscle and uterus when given to a rat. He called the substance enteramine.

A few years later, in 1945, Irvine Page was working at the Cleveland Clinic, Ohio, studying the cardiovascular system. He was trying to find a specific vasoconstricting substance in the blood, which he thought was to blame for high blood pressure. But whenever the blood began to coagulate, another substance was produced which also constricted blood vessels, and it was interfering with his research. So, with biochemist Arda Green and organic chemist Maurice Rapport, he set about trying to isolate and remove the offending substance – the same 'adrenaline mimicking substance' discovered decades earlier.

Green's biochemistry background allowed her to develop a method to test substances isolated from the blood, using the artery of a rabbit's ear to see if the substances caused constriction of the blood vessels. Rapport spent his days isolating compounds from hundreds of litres of blood, collected from the Cleveland slaughterhouse. First, the blood had to be passed through a cheesecloth as it coagulated, to produce serum. Then, Rapport developed a multi-step process which, eventually, allowed him to isolate this vasoconstricting substance as pale yellow crystals – just a few milligrams from 900 litres of serum. In 1948, they published their findings, calling these precious crystals 'serotonin'. A year later, now at Columbia University, Rapport identified the chemical structure of serotonin as 5-hydroxytryptamine, or 5-HT.

Over at Harvard University, Betty Twarog was the PhD student of John Welsh, an expert on invertebrate neurobiology. They had identified the neurotransmitter which helped mussels to hang on to rocks so tightly, but were looking for the one that allowed them to release. Reading a paper from the team at the Cleveland clinic, Twarog became convinced serotonin could be that neurotransmitter. Soon, experiments proved she was correct. Twarog wrote a paper laying out the role of serotonin as a neurotransmitter in invertebrates – a paper which wasn't published until two years later, in 1954, because the *Journal of Cellular and Comparative Physiology* 'Had not bothered to review a paper on an unknown neurotransmitter by an unknown author.'

Twarog moved to the Cleveland Clinic in 1952, where she worked with Page. Convinced that serotonin would also be found in the brains of vertebrates, she persuaded a dubious Page to support her work in the area. Sure enough, in 1953, she discovered serotonin in the mammal brain. We now know that serotonin in our brain is vital for a whole range of functions, from sleep to love. But the aspect it is most commonly associated with is mood. Too little, it is said, and we feel low. Boost it, and happiness returns. However, this statement is more controversial than it first appears, because studies just haven't supported a clear one-to-one relationship between mood and serotonin levels.

The happiness hormone?

So where did this idea, that serotonin and depression were linked, come from in the first place? Well, like many big breakthroughs in science, this theory was precipitated by a series of chance discoveries. The first of these occurred in India. Scientists discovered that the dried root of *Rauvolfia serpentina* (also known as Indian snakeroot), used in traditional medicine, could lower blood pressure. There were also hints that it might have some effects on mental disorders. This piqued the interest of Western psychiatrists, who ran clinical trials in people with schizophrenia, using a substance isolated from the plant. Sadly, they found this drug, called reserpine, had no effect on the illness directly, but did note that the patients seemed calmer. There were also anecdotal reports of benefits to anxiety, compulsions and difficulties with motivation.

However, there were also reports of a dark side to this drug, and in the late 1950s, after an initial surge in use, it was replaced by safer options. One of the biggest reasons for this was the concern that the drug could cause depression in some people. It was discovered that reserpine depletes the brain of serotonin, and also other neurotransmitters like dopamine and noradrenaline, so scientists began to wonder if this was the reason for its side effects.

Around the same time, it was found that a tuberculosis drug also affected patients' moods. Treated individuals regained their appetites, felt happier and less apathetic and even, sometimes, experienced euphoria. Of course, if you had a horrible illness and suddenly started feeling better, it wouldn't be surprising if your mood improved too. But other side effects like drowsiness and a dry mouth suggested the drug might be having an impact on the nervous system. So the drug was tested in psychiatric patients, and soon evidence started to emerge that iproniazid (as it was called) might be effective in treating depression.

Just like reserpine, iproniazid had side effects that meant its popularity quickly declined, but its mechanism of action was a vital ingredient in the understanding of depression. In 1952, scientists found that the drug blocked an enzyme which usually breaks down the monoamines (a class of neurotransmitters including dopamine, noradrenaline, adrenaline and serotonin, amongst others), increasing levels of these chemicals in the neuron, ready for release into the synapse.

Taken together, the effects of these two drugs and their mechanisms of action suggested a simple hypothesis. If you reduce the levels of monoamines in the brain, you produce depression. If you boost them, you make people feel better. And so the hypothesis was born. Over the following years, this concept gathered strength, as new antidepressant drugs were developed, and their mechanisms of action seemed to fit.

Tricyclic antidepressants, for example, were discovered when searching for better antipsychotic drugs. These have a complex mechanism of action, but do affect both serotonin and noradrenaline levels, so could provide support for the theory.

In the 1960s, research began to hint at the idea that it was serotonin specifically that was linked to depression; for example, people with depression who had died by suicide seemed to have lower concentrations of the chemical in parts of their brains. Pharmaceutical company Eli Lilly began to search for ways to boost serotonin in the brain without affecting other neurotransmitters. In 1974, they were successful, publishing a report on their first selective serotonin reuptake inhibitor (SSRI): fluoxetine. By 1988, fluoxetine had been approved by the FDA and released under the trade name Prozac®.

SSRIs are now the most common class of antidepressants. These drugs revolutionised the treatment of mood disorders because of that first S – they are selective. This means they only target serotonin, and so have fewer side effects than drugs which affect multiple neurotransmitters.

The mechanism of SSRIs sounds, on the surface at least, relatively simple. Just like cocaine does for dopamine (see Chapter 3), SSRIs block the serotonin transporter. This prevents excess serotonin being sucked back up into the neuron that released it, meaning it hangs around in the synapse a lot longer after it is released. This, in theory at least, produces an antidepressant effect.

The success of these drugs seemed to cement the idea that serotonin was linked to happiness, and that by boosting it, depression could be conquered. And if boosting levels of this neurotransmitter improved depressive symptoms, surely that means that the problem was caused by too little being produced? The theory of serotonin as a 'happiness chemical' was born.

But not all the evidence stacks up here. Direct evidence that people with depression have lowered levels of serotonin in the brain is hard to come by. Some studies *have* found low levels of serotonin and linked chemicals in the blood and cerebrospinal fluid of people with depression, but others haven't. And we have to interpret these results carefully, as these tests don't necessarily match up with levels in the brain.

To test the theory, researchers turned to experiments that deplete levels of monoamines in the brain. Over the years, a large number of studies have found that doing this in healthy subjects *doesn't* produce depression or low mood. In fact, it doesn't even increase symptoms in everyone who has depression.* So the idea that low serotonin means low mood just isn't right – there must be something else going on.

When it comes to treatment, while some people with depression do respond to serotonin-boosting antidepressants, there is a proportion of patients who don't. And even in the people who do get better, there is a problem for the theory. When you take an SSRI, it blocks the transporter almost immediately, so serotonin levels should increase very quickly. But the improvement in mood, if it happens, doesn't occur for weeks – often six weeks or more. If boosting serotonin makes you happy, and SSRIs boost serotonin so quickly, why don't people feel better straight away?

Newer theories argue that while serotonin does have a role in depression, it's not the direct relationship between levels and mood initially hypothesised. To understand how this might work, we need to look in a bit more detail at the anatomy of a serotonin neuron.

The off switch

Like many neurotransmitters, serotonin can have different effects depending on where in the brain it is released, and what neuron it is received by. Part of this is because there are different types of serotonin receptors,

* Although some studies have found depletion may cause relapses in people who have recently recovered from depression, and who are on SSRI drugs.

which have different effects. As well as receptors on the ends of neurons, ready to receive serotonin from the previous cell in the chain, there are what are known as 'autoreceptors'. These are found on the first neuron, the one that is releasing the serotonin. They can be on the cell body, the dendrites, the axon or even on the first side of the synapse, where they are known as presynaptic. These receptors have the important job of making sure serotonin levels don't become dangerously high (like all neurotransmitters, too much serotonin can have nasty side effects, for example psychosis), using a negative feedback loop.

Let's walk through this process. An electrical signal is triggered in the first neuron and travels along to the synapse, where serotonin is released. Much of the serotonin travels across the synapse to the post-synaptic receptors, some is sucked up by the transporters, and a little bit diffuses back to the autoreceptors, but not enough to activate them. The second neuron has received the signal and is able to send its message onwards.

Now let's imagine there is an excess of serotonin released, either through repeated activation of the presynaptic neuron or through some type of anomaly in the system. Serotonin will, as before, travel across the synapse to activate the post-synaptic neuron. But it will also travel back, and as there is more of it this time, it will activate the autoreceptors on the presynaptic neuron. These will then damp down activity of that neuron, preventing it from firing, so it can't release any more serotonin. As serotonin levels begin to decrease (the transporters suck it up, and no more is released), the autoreceptors stop their inhibition, and the neuron is able to fire again. It is an elegant way for the brain to ensure that levels of serotonin don't rise above a certain threshold.

You can imagine the system as a bit like a high-tech smart bath. In my (imaginary) bath, rather than a simple overflow drain, there is a sensor which detects when the water level is getting too high and turns off the tap. This means you can never overfill the bath, and avoids wasting water or causing a flood. If you took the plug out, and the water level began to drop, the tap would start again, ensuring the perfect level of water. The sensor is playing a similar role to the autoreceptors, which turn off the release of serotonin when it gets too much, then allow it to flow again when levels drop.

But what if something goes wrong with these autoreceptors? J. John Mann from Columbia University has been accumulating evidence that suggests this might be the problem in depression. Doctors still tell people with depression that serotonin levels in their brains are low, and that SSRIs correct them, and Mann was interested in what exactly this meant. So in the 1980s he began to look at the brains of people with depression who had died by suicide, and compared these to the brains of people who had died without depression. And he found something surprising. 'They don't have a deficit of serotonin neurons,' he told me. 'They don't even have a deficit in the enzyme that makes serotonin. Actually they have the opposite – they have more neurons and they have more enzyme... and even when we measured the level of serotonin right inside the neuron, in the cell bodies, they had more.'

Mann began to look at the serotonin receptors, to see if there was something going on there. He found that in the post-mortem brains of people with depression, there were more serotonin autoreceptors, and following up with brain-imaging studies in living patients confirmed these findings. Interestingly, even when people had recovered from a period of depression, this high level of autoreceptors remained. In fact, his team found that autoreceptor number could be predictive. They studied the children of people with depression, who were all well at the time, but were at higher risk for developing the disease, and found that those with the biggest increase in autoreceptors were most likely to go on to develop problems with depression in the future. It seems this might be the mechanism for the heritability of depression: if you are born with more autoreceptors, you are at greater risk. But you can also develop more of them throughout your life, and it is likely to be this mix of genes and environment that triggers depression in most people.

But all this didn't explain *why* there was a link between having more of this type of receptor and depression, so to understand the mechanism more fully, Mann turned to studies on mice. Working with mice genetically engineered to have a high number of autoreceptors, Mann and his colleagues confirmed what was going on. In these animals, only a small amount of serotonin needed to be released in order for the autoreceptors to be triggered, shutting down that neuron and preventing any more serotonin emission.

It's as if our smart bath had accidentally been made with many of the auto-shut-off sensors, dotted all around the tub. It wouldn't matter how much hot water was waiting in the tank, every time you tried to run a bath the sensors would immediately turn the tap off, so your bath would never fill up.

Coming back to the mice, Mann found they had lower firing of serotonin neurons and less release of the neurotransmitter. They were also more prone to the mouse version of depression.* He believes the increase in serotonin inside the cells is the brain trying to compensate for this lack of release, but one that does no good. As he puts it: 'The bottom line is, the autoreceptor keeps shutting off the firing, so they effectively have a serotonin deficiency.'

This theory is an interesting one, as it explains one of the big mysteries of depression treatment: why SSRIs take so long to work. Studies in rats have found that SSRIs initially reduce the firing of serotonin neurons, which fits with Mann's findings. SSRIs block the reuptake of serotonin, so there is more in the synapse, and more will diffuse back to activate the autoreceptors, shutting down the neurons faster. But just like a drug user whose brain fights back in response to high levels of their drug of choice (see Chapter 3), boosting serotonin causes the brain to fight back. Over weeks of dosing, the

* You can test whether a mouse is depressed in various ways. They might freeze for longer when presented with a tone that had previously been linked with a shock. If placed in water, they might just float, waiting to be rescued, rather than trying to escape. And they might lose their preference for sweet tastes or even rewarding brain stimulation. Anxiety can be measured too: if offered the choice of a closed or an open arm of a maze, anxious animals will spend more time in the closed areas, not exploring as much as non-anxious animals.

autoreceptors become less sensitive, and even begin to decline in number. This allows the neurons to begin firing again, and serotonin release increases. A study in humans in 2013 found that over seven weeks of SSRI treatment, autoreceptors decreased by 18 per cent.

So could it be that the serotonin hypothesis was right after all? That these people did have a reduction in serotonin release, but blocking reuptake wasn't enough; instead this slower mechanism is needed before levels rise and they start feeling better? It certainly seems this might be one piece of the puzzle. This receptor might also be important for mood in the healthy population. Studies have found that people with a certain version of a gene, which causes changes in these autoreceptors, are more prone to poor self-esteem and neuroticism, and tend to be more introverted socially and have fewer friends. They are also more at risk of depression. Maybe serotonin really is a 'happy hormone' after all!

Always look on the bright side of life

While neuroscientists and pharmacologists have been working on *how* SSRIs change serotonin levels in the brain, psychologists and psychiatrists have been mainly focused on *whether* they produce an improvement in symptoms. But this leaves a missing link in the puzzle, the *why* question: why do SSRIs help some people to feel better? This gap highlights one of the biggest issues in psychology and neuroscience research, and a barrier I have repeatedly run into while writing this book, because translating molecular changes into behaviour, and understanding how one can cause the other, is extremely tricky. But it is a challenge that Catherine Harmer, director of the Psychopharmacology and Emotional Research Lab (PERL) based at the University of Oxford, wasn't going to shy away from, as she told me:

> Most research at the time I started didn't really cross those different divisions… and those different research folk didn't really communicate very much with one another. And I just became really interested in how these different things related to one an other, so how SSRIs, or antidepressants more generally, actually led to any improvements in the symptoms of depression… Why does increasing levels of serotonin in the brain allow people to feel better? It seemed like there was a missing gap there in explanation.

To figure this out, Harmer looked at the effects of SSRIs in healthy volunteers and people with depression, to see how they change their view of the world. But first, she needed to see if there were differences between the two groups before they were given the drugs. Her group found that when shown pictures of people's faces, healthy participants will tend to focus mainly on those with happy expressions, while people with depression, even when they are in remission, spend more time focusing on the sad. They also interpret faces differently: when shown an ambiguous expression, healthy controls tend to see it as happy, while people with depression are equally likely to think it is sad or happy. The same is found in other areas, so people with depression are more likely to remember the negative words on a list, for example, while controls remember the positives.

What her findings add up to is a negative bias in people with depression. Someone with depression is more likely to notice and remember the bad things they experience, rather than the good. And this may be the cause of their low mood, or at least help to maintain it. It may also encourage them to withdraw from society, because if you experience most interactions with other people as negative, you are less likely to repeat them. This can then reinforce the problem as their growing isolation makes their depression worse.

Next, working with clinical researcher Beata Godlewska, Harmer gave her subjects with depression SSRIs, boosting the levels of serotonin in their brains, and gave them the same tasks again. And she noticed an immediate difference. The group with depression were more able to recognise happy facial expressions and found it easier to remember the positive words. Their negative bias had shifted.

What is particularly interesting about Harmer's results is how quickly the changes in processing appear. In some cases, they are visible after a single treatment, and certainly well before the individual has noticed any effect of the antidepressant (remember it often takes six weeks or so before people start feeling better). That means that this could be the *reason* their symptoms begin to improve. Harmer explains:

> Increasing levels of serotonin doesn't boost up mood, but what it does is make you collect information from your environment in a more positive way. But that doesn't immediately make you feel better... They don't notice it at a subjective level, so they don't report feeling any more positive, but still they're collecting more positive information and then over time and experience and

> interaction with things that are going on in your life, you'd expect that to allow you to feel better. So it suggests that serotonin doesn't influence mood directly, it affects emotional processing, which then impacts mood... You need time to learn from that change.

This is where the amygdala activation is key. The amygdala responds to anything in our environment that could be important for our survival. This can be positive things, but often it is negative – something we should fear, for example, like that large (imaginary) shark. Harmer and Godlewska also imaged the brains of their participants, and found that in the people with depression, seven days of SSRI treatment reduced the activation of the amygdala in response to fearful faces. With serotonin-boosting drugs, it seems that the amygdala becomes less sensitive to negative events, allowing people with depression to focus more on the positives. This fits with the idea that an imbalance between the reactive limbic system and the more reasoned prefrontal regions, which can inhibit our emotional responses, is involved in mood disorders.

So it seems that over the long term, emotions, controlled by our limbic system, can impact on our moods. If we focus on the good things in life, we feel happier, but if our brains are in a state where we are forced to focus on the negatives, mood disorders can develop. This fits with a theory that suggests our mood is based on a combination of our expectations and our experiences. If good things have happened to us, and we expect more good things, we will be in a good mood. The more bad experiences we have, and the more we expect bad things, the worse we feel. This is an interesting idea because it explains not just mood in healthy people, but also in depression – the more bad things that happen, the more you are at risk of developing depression. But it's not a guarantee, because individuals may react differently, depending on how they think about the future.

Interestingly, there also seems to be self-reinforcing, as people experience events as more positive when they are in a good mood. This could explain why depression is so hard to beat. Just as in Harmer's studies, once your mood is low, you are likely to experience neutral events as negative, and this will make you feel worse. And while most of us can drag ourselves out of a funk, when we occasionally end up in one, something in the brains of people at risk for depression seems to make this harder.

SSRIs might be a way to help people out of this spiral, but, as we know, they don't work for everyone. Harmer found that not everyone has this change in processing after the first dose or two of antidepressant, and those that don't tend to be the people who fail to respond to the drug in the longer term. This could be an extremely important finding clinically. Currently, while doctors know that only around a third of their patients will respond to the first antidepressant they try, there is no way to know who will and who won't until they have been taking the drug for six weeks. If that doesn't work, they try the next… and wait another six weeks. The cycle continues until they find one that works or run out of options. This means it can often take several months of trial and error before effective treatments are found, and this is a long time for someone with severe depression to wait. Harmer's findings offer a tantalising glimpse into how we could test people after just one or two doses, using a test like those used in her research, and know immediately whether that drug would work for them. This would mean we could find the correct treatment much faster, and get the patients on the road to recovery a lot more rapidly. And that could significantly change their quality of life.

This still doesn't explain *why* some people with depression don't respond to SSRI treatments, but Harmer believes there are several plausible explanations. First, she sees the environment as an important factor in recovery. 'When you're depressed you become very isolated... so even if you are starting to be able to pick up more positive messages from other people, if you're not seeing people you can't pick them up. Or if you have an extremely toxic, negative environment, there might not be anything positive to pick up.'

Someone on SSRIs might be more able to pick up the positives in a social encounter, rather than the negatives they would have seen before. Over time, this could encourage them to repeat that encounter, helping to pull them out of the depressive spiral. But if the only social interactions they have are extremely negative (perhaps they are in an abusive relationship, or a toxic work environment), there would be no positives to detect, so all the serotonin in the world couldn't help them find the good in the situation. This is why treating depression is a multi-step process, and why environmental changes can be as important as drug therapy. Cognitive behavioural therapy can also have a role here, in helping people to challenge and change their way of thinking, and this might explain why drugs and therapy together are more successful than either treatment alone.

Another explanation for the failure of serotonin-boosting drugs in some people with depression is a simple one: they might not have low serotonin. Harmer believes that depression can occur for a number of different reasons, and each might have a different biological basis. So, trying to treat everyone with SSRIs is like giving everyone who has a runny nose anti-allergy medication. Yes, it might help those whose runny nose is caused by hay fever, but it won't do a thing if the underlying cause is a cold virus. Finding that underlying cause will allow doctors to treat patients in a more tailored and hopefully more efficient way.

Hope on the horizon

Of course, there are other theories for why SSRIs take so long to work. One suggests serotonin might be acting not as a neurotransmitter, but as a growth factor, encouraging the development of new neurons in the brain, which is known as neurogenesis. We know that chronic stress in mice leads to a loss of neurons in a region of the hippocampus, and that SSRIs can protect against this loss. And in humans, the same area of the hippocampus is smaller in people with depression who are left untreated, but not in those who have been treated with antidepressants. This suggests that the drugs are preventing neurons from dying off, or even increasing production.

The hippocampus is involved in connecting emotions with memories, so if this area shrinks, this could lead to emotional problems. It may be that the delayed benefit of serotonin-boosting drugs comes from its role boosting neurogenesis and restoring the function of the hippocampus. Or, as is probably more likely in the incredibly complex system that is our brain, it could be a combination of both.

Other brain chemicals involved in neurogenesis have also been implicated in depression. One of these is Brain Derived Neurotrophic Factor (BDNF), a protein which promotes neuron growth. Levels of BDNF have been found to be lower in people with depression. Treatment with SSRIs restores levels to normal, and exercise can also boost levels, which could explain its antidepressant effects.

This leads to another question: could we help people with depression recover faster by targeting neurogenesis directly? This seems to be what treatment with ketamine does, which makes it a really exciting area of research. Better known as an anaesthetic used by vets, or a party drug, ketamine acts on glutamate receptors, blocking one particular receptor that seems to be responsible for the toxic effects of too much glutamate in the brain. At the same time, it promotes the release of more glutamate. Unable to bind to the blocked receptor, this excess of the excitatory neurotransmitter binds to a different receptor, boosting neurogenesis in the hippocampus. Amazingly, ketamine can improve the mood of people experiencing severe depression in just two hours. However, it can have side effects, including memory loss, and at higher doses it is abused by some people. It also has the potential for addiction (see Chapter 3), so more studies are needed to prove it is not just effective but also safe, and non-addictive when used in clinical settings.

GABA (gamma aminobutyric acid, which we met in Chapter 2) may also play a role in depression. This brain chemical reduces neuron activity, and too little of it, compared to the level of glutamate, may lead to anxiety. Ketamine raises the levels of this chemical, as do anti-epileptic drugs, which may also be beneficial in depression. And in healthy people yoga has been found to increase levels of GABA in the thalamus, and this correlates with reductions in stress and improvements in mood.

Another factor that seems to play a role in depression is inflammation. As we saw in Chapter 3, if levels of messenger molecules called proinflammatory cytokines are elevated in the blood, they can cross the barrier between the blood and the brain, inducing 'sickness behaviour'. This makes sense if you have an injury or infection, but if it persists long term, it stops being beneficial. In depression, high levels of cytokines may persist, causing people to get stuck in this sickness behaviour. They continue sleeping a lot, avoiding normal social interactions and not enjoying activities they previously liked, all symptoms of both acute illness and depression.

Studies have found that there are higher levels of inflammatory markers in the blood and the cerebrospinal fluid of people with depression, and in the brains of those who have died by suicide. These high levels of cytokines have only been found in about a quarter of otherwise healthy people with depression, but interestingly they are more common in those who haven't responded to SSRIs and other antidepressant treatments. There is also more direct evidence. If people are given a drug that increases inflammation, they are more at risk of depression. Inhibiting cytokines in people with autoimmune disorders (which are associated with widespread inflammation) can reduce their depressive symptoms.

From a biochemical point of view, there is some evidence that high levels of inflammation encourage the breakdown of a molecule called tryptophan, which is needed for the body to make serotonin. This could explain why SSRIs don't work in this population – blocking reuptake of the chemical won't have much effect if there isn't much of it around in the first place. There is also the link with dopamine that John Salamone has been investigating (see Chapter 3), so perhaps these people with depression who have inflammation need help boosting their dopamine, not their serotonin levels.

Life's balancing act

It seems to me that we might be causing problems by thinking of depression as one disease, and so looking for a single cause. This works fine for something like diphtheria. Once we discovered it was caused by bacteria, we could develop antibiotics to treat it and vaccines to prevent it. But I wonder if depression is a symptom, rather than a disease, so more like pain. If you go to the doctor with a pain in your leg, yes, they can give you painkillers to treat the symptom. And these might work. But that doesn't mean that too little morphine was the cause of your pain. Your leg might have been broken. Or maybe you pulled a muscle. Or perhaps there is a huge shard of glass sticking out of your shin.

Similarly, it might be that there are distinct types of depression, one mediated by low serotonin release, one by inflammation and one by an overactive stress response. Until we have a way of distinguishing which of these is the problem for an individual patient, it's going to be challenging to develop treatments that address the cause, rather than just treating the symptom. This is something scientists are working on, and early results show promise. For example, a recent study led by Annamaria Cattaneo, from King's College London, found they could identify how likely patients were to respond to conventional antidepressant treatment by carrying out a blood test for inflammation levels. But these findings haven't yet translated into general use.

The more we learn about the brain, the more we realise just how interconnected it is, and how a change in one place can have knock-on effects elsewhere. So another idea is that all these explanations are valid, and any combination of them can be enough to trigger depression. But it is likely they are also all interrelated. We know, for example, that cortisol, released when we are stressed, promotes inflammation. It also acts to lower serotonin levels, which can then lead to changes in other neurotransmitter systems. And this can have knock-on effects on neurogenesis, alongside cortisol's direct neurotoxicity. Inflammation itself can trigger more stress, creating a powerful feedback loop.

By looking at the complexity of depression as a multi-faceted condition, we can also explain some of the findings in psychology. Depression is often triggered by life events, but something that might cause an episode in person A can be handled without issue by person B. This difference could be down to a change in any point in the web. Perhaps person A has a very negative outlook on the world, so perceives something as more threatening, or has an overactive stress response, caused by trauma in early life. Either (or both) of these would mean they produce more cortisol, which could be enough to trigger the cascade and cause depression. Or perhaps person B is lucky enough to have a genetic variant that means they have low numbers of serotonin autoreceptors, so they are affected less by changes in the release of the neurotransmitter. Or maybe person A is suffering from chronic inflammation caused by lifestyle factors or illness. Whatever the reason, or combination of reasons, it adds up to the same outcome: depression.

But can learning about these disorders help us better understand mood in the general population? It's becoming more popular to think about psychiatric conditions as one end of a scale. Take autism, for example; once it was thought of as a distinct disease, but we now think that everyone falls somewhere along a spectrum when it comes to the traits associated with autism. If you fall far enough along the spectrum that it interferes with your daily life, you are diagnosed with the disorder. The same is true for schizophrenia. Some of the general population score higher on measures of schizotypy, such as having unusual beliefs, but it isn't until this becomes a problem that the diagnosis is made. So could depression be similar? Are we all somewhere on a depression scale, and could the same mechanisms that underlie depression underlie the differences in outlook within the healthy population? This is certainly an appealing idea. We all know some people who seem to bounce back instantly from hard times, and others who are affected dramatically by the tiniest stumbling block. I asked Harmer whether this could be the case:

> Depression is often thought about as falling on a spectrum – it is normal to have some symptoms but it is only when

> they come together at a more severe level and have a negative impact on your life that you reach the 'criteria' for an episode. It's also true that we will have different levels of vulnerability to depression, but this might only become depression in conjunction with other things going on in your life – perhaps life stress, inflammation or illness.

She also believes her work on people with depression could apply to the general population: 'I think of cognitive processing as a vulnerability factor for depression – we all have different levels of positive and negative bias, which may get exaggerated or lead to a negative effect in combination with stress, for example.' If this is the case for Harmer's work, it seems reasonable to think it could apply to the rest of the studies we have discussed in this chapter. Changes at any point in that depression network could contribute to making some of us more resilient, and others more prone to low mood, even if not clinically depressed. This complex set of interacting traits might explain why our mood is so easily influenced by all sorts of things, from the weather to how well we slept. Our brains are carrying out a careful balancing act, and small changes can have major consequences for some people.

Thinking back over what we have learnt throughout this chapter, I wonder whether now, knowing what I know about how my brain and body is responding, I could swim with sharks without feeling that surge of fear. And, perhaps more importantly, are there science-led techniques we could all use to better control our emotions and lead happier and healthier lives? There are a few techniques I discovered while writing about stress, and it turns out that in a lot of cases, managing stress in everyday life has knock-on impacts on long-term happiness.

First, it is vital to make sure you get enough sleep. As we will see in Chapter 5, lack of sleep increases your emotional reactivity, having a negative impact on mood, as well as cognitive function. Relaxation techniques can be great for improving sleep and mood. Deep breathing has been shown to reduce cortisol levels and heart rate, while meditation, yoga or tai chi, which tend to focus attention on your breathing and the present moment can also be effective in reducing stress and improving mood.

One mood booster that has long been known is exercise. We don't know exactly how it has its effect, but animal studies have shown exercise boosts growth factors like BDNF, which may help keep the brain healthy. It also changes the levels of a host of other brain chemicals, including serotonin and dopamine, which may contribute to its mood-boosting effects.* Another helpful technique is gratitude. Whether it's keeping a journal, practising loving kindness, meditation or just counting your blessings at the end of a long day, there is growing evidence that focusing on the positives in your life might have benefits, for health and mood. But practising is a key word here. Many of these techniques don't work immediately; however, over time, you can teach your brain to focus on the positives and become more resilient to stress.

---

* Interestingly, the evidence for an increase in brain endorphins causing the 'runner's high' is limited, despite the idea being widespread. They may well play a role, but there aren't enough studies showing a direct link to say for sure that these opioids create the good mood many feel after exercise.

CHAPTER FIVE

Sleep, the Brain's Greatest Mystery?

There are a number of things a human being needs to be able to function: food and water, shelter, warmth and, perhaps most importantly, a strong WiFi connection. But one need that is just as fundamental, but often overlooked, is sleep. Sleep is so vital that sleep deprivation has been used as a form of torture for hundreds of years.* And yet many of us go through life sleep deprived, without realising the toll it can take on our bodies.

The effects of acute sleep deprivation are fairly obvious. If you have ever pulled an all-nighter to finish an essay due the next day, been up for hours with a crying baby or stayed at a party a little too long when you had work in the morning, you will have felt the effects. The day after is probably a blur – you may have felt grouchy or tearful, and struggled to complete simple tasks or remember information you can normally recall easily. You probably felt hungrier than usual, reaching for sugary snacks and caffeine to try to give yourself a lift. And you will definitely have felt tired, perhaps noticing your eyelids drooping and finding yourself yawning more than usual.

* And it's not just a historical method. Sleep deprivation was one of the 'enhanced interrogation techniques' used by the US military on prisoners in Guantanamo Bay and elsewhere.

All these effects are down to chemical changes in your brain that occur when you don't get enough sleep.

But what about long-term sleep deprivation? The average adult needs seven to nine hours' sleep a night – if you consistently get less than this, you are said to be in 'sleep debt'. If it's not repaid, over weeks or months even mild sleep debts can build up and start to impact your daily functioning. In fact, habitually not getting enough sleep can put you more at risk of illness by damping down your immune system.

So what about those people who claim they can get by just fine on four or six hours a night? The likelihood is they are deluding themselves. Ying-Hui Fu, at the University of California, and her team recently discovered some genetic variants that *do* allow people to function well on very little sleep. Unfortunately they are very rare. So rare that you are actually more likely to be struck by lightning than to have one of these genes.

Telling someone this is all very well, but sleep isn't an easy option for everyone. I have never been a particularly good sleeper. As a baby, I wasn't fond of naps, and my parents would often have to drive me around in the car for hours to get me to nod off. Throughout my childhood, I slept fine, as long as the conditions were right. In a quiet bedroom, with blackout curtains and my own bed, I could drop off, no problem. But when it came to sleepovers, it was a different story. Friend after friend would be overtaken by slumber around me, often in the weirdest of positions, while I lay there. A ticking clock, a friend's breathing, discomfort from lying on a thin mat on the floor – any one of these was enough to stop me from sleeping.

This is still a problem I have today, and it has always fascinated me that something of such vital importance to our long-term health can come so easily to some people, and be a never-ending battle for others. So I wanted to find out what might be going on in our brains that allows some people to enter the land of nod so easily, while I have to use every trick in the book to coax my brain into letting go of consciousness. The first step is to look at why we (mostly) get sleepy at night, which means exploring our built-in body clock.

Keeping time

We have body clocks, or circadian rhythms, throughout our body, controlling the timing of processes like the release of digestive enzymes from the liver and the energy production cycle of our cells' mitochondria. But these are all kept on schedule by a master clock in our brain, found in the suprachiasmatic nucleus (SCN), part of the hypothalamus. We don't understand the exact mechanism by which the brain synchronises the rest of the body, but recent research suggests it's not just the neurons in the SCN that are important. Astrocytes, brain cells previously thought to be there just to provide support and keep the neurons healthy, actually have the ability to control circadian rhythms, even when there are no neurons present. This suggests they might have a more important role than we previously thought.

Interestingly, our internal timekeeping isn't perfectly synced to the length of a day. If an animal is left to rely just on its body clock, it often won't stick exactly to a 24-hour cycle.[*] This phenomenon is known as free-running and is achieved by taking away anything that could give the animal a clue as to what time it is. Lighting, temperature and food availability must be constant, and the animal must be kept in these conditions for several days. This, of course, makes it challenging to do studies on humans, but those that have been done tell us that the same can happen to us. Volunteers quickly find themselves drifting away from outside time and may end up hours out of sync in a matter of days. In fact, studies have found that most humans can stretch their body clocks to run around half an hour shorter or longer than the usual 24 hours.

Luckily, most of the time we don't have to rely just on internal time-keeping. Our body clocks have a reset switch, which is controlled by light. During the day, bright light is detected by special cells in our eyes, which send signals to the superchiasmatic nucleus. This initiates a cascade of changes that make us feel awake and alert. Later in the day, as the light begins to dim, this information too is carried to the SCN, which again initiates changes, this time to make us feel sleepy. The timing of the cycle we run on can be changed by adjusting the timing of light and dark

This light entrainment is useful, as it allows us to (gradually, at least) adapt to different time zones when we travel. But the modern world can cause problems, because we are surrounded by artificial light, which can wreak havoc on our delicate circadian rhythms. To understand

---

[*] The fact that it isn't precisely 24 hours is actually where the term circadian came from – it's a combination of the Latin words circa (approximately) and diem (day).

why, we need to look at how the resetting works in a world where the only light comes from the sun.

In this world, you wake up at dawn. Soon, the sun is high in the sky, producing lots of light. This light signals the SCN and we feel awake. In the evening, the light gets dimmer as the sun sinks down towards the horizon. This, along with the time that has passed since you first saw the bright light, signals the SCN to release a hormone called melatonin, making you feel sleepy. But there is another element to this process: the colour of the light. Morning light has a large blue component,* and blue light, with its short wavelength, is particularly good at resetting the circadian rhythm by delaying the release of melatonin.

But humans are always inventing new things, and this is where the problems arise. We no longer live in a world where all, or even most, of our light comes from the sun. In fact, many of our lights are now LEDs, designed to replicate blue-ish daylight. Our phones, laptops and tablets also give out a lot of bright light, tinged with blue. This means that as we spend our evenings on social media or watching YouTube videos of hilarious cats, our eyes detect this light, and send signals to our SCN. And our poor, confused SCN, believing it is morning, delays its release of melatonin. Eventually, we drag ourselves to bed, but with less melatonin we may struggle to sleep, perhaps checking our phones every now and again, making the problem worse.

So what can we do? If you do struggle to drop off in the evening, it can be helpful to get some light first thing

* This is because of the height of the sun in the sky, along with the temperature and the amount of dust in the air.

in the morning, by going for a walk, for example (being outside, even on an overcast day, provides a surprisingly greater amount of light than even sitting by a window indoors). If that's not possible, it might help to invest in an SAD lamp. These are designed to help people with seasonal depression (also known as Seasonal Affective Disorder) and provide bright daylight-like light to help set your circadian rhythm. The other thing is to reduce light exposure in the evenings, especially blue light. If turning off your devices isn't an option, many now have built-in blue-light filters you can set to come on at sunset, or you can download apps that do the same. And dimming the screen as much as possible is also a good idea.

Understanding how our brains react to light can also help with jet lag, something I suffered with very badly on a trip from the UK to Singapore last year. Lying in my huge hotel bed, I remember glancing at the clock. The red lights flashed back, mockingly; 04.16, they said. I sighed and stared at the dark ceiling, my mind whirring. I had done everything right – I set my watch to Singapore time the moment I got on the plane, ate a light dinner at 7 p.m., took a relaxing bath, dimmed the lights, but despite my best efforts, jet lag was winning. My body knew that it was only 8 p.m. at home, and was asking for some chocolate and an episode of *Masterchef*, not sleep. No matter what I tried, I just couldn't seem to override it.

This is an experience most of us will have had at some point in our lives, and it often feels like fighting jet lag is a losing game. So much so that when a friend stayed with us recently between work trips, she decided to stay roughly on Japanese time for the two weeks she was in the UK.* Luckily, the science of body clocks can step in to help out.

One way scientists keep track of someone's body clock is by measuring their core body temperature, which varies by up to half a degree (centigrade) depending on the time of the day. The minimum temperature happens a few hours before you normally wake. When I arrived in Singapore, my body clock was still on UK time. It wanted to go to bed at 10 p.m. in the UK, and, had I measured it,† my minimum temperature would have occurred at around 4 a.m. UK time. But that means I wasn't getting sleepy until 6 a.m. Singapore time, just as everything around me was coming to life.

We can help the body fight this by providing light at the right time. I tried this on a more recent trip to Singapore. To shift my body clock earlier, so it matched up with Singapore time, I needed to get light exposure only *after* that minimum temperature point, known as Tmin; studies have shown that the few hours after Tmin are the sensitive period where light can most effectively shift our circadian rhythms. On the day I arrived, my Tmin would have occurred at midday in Singapore (4 a.m. in the UK). I made sure I stayed in a dark room for the morning, keeping the curtains closed and wearing sunglasses to breakfast.‡ As soon as midday hit, I opened the curtains

* Something that I wasn't too happy about when the sound of the shower running woke me up at 3 a.m.!

† I didn't, of course. Taking your temperature at regular intervals is unlikely to help you overcome jet-lag-induced sleep problems!

‡ The things I do for science! I do wonder if the other hotel guests thought I was some kind of rock-star wannabe, or if they just assumed I was hungover?

and spent some time outside, to get as much bright light as I could. Using this technique, the body clock is meant to shift about an hour a day.* So, on the second day, I could open the curtains at 11 a.m., then 10 a.m., until I was back in sync. I am only one person, and it was only one experiment, so not particularly scientific, but I think it worked. I certainly didn't struggle as much to fall asleep as I had the previous trip, so I'm inclined to think it was helpful, and I would definitely try it again.

Travelling in the opposite direction, our bodies have a different problem. When I returned to the UK from Singapore, I had to try to stay up long enough to go to bed at 10 p.m. UK time, while my poor body thought it was 6 a.m., and wanted to have been asleep hours ago. To match these timings up, I needed to delay my body clock. To do this, I needed light before my Tmin – so in the early evening. Luckily, travelling in this direction, natural light tends to fit with the timings we need to alter the body clock, so I didn't have to do anything in particular. As I experienced daylight before my Tmin, this naturally helped to shift my body clock, which is one of the reasons it seems to be easier to adapt to westward than eastward travel. Another is that it is simply easier to force yourself to stay awake, by chatting to people, watching TV, or drinking coffee, than it is to force yourself to sleep when your body really doesn't want to.

Setting the clock

But what about places in the world that don't have a regular light-dark rhythm for people to set their body

---

* Although, despite this number being quoted all over the place, there is little scientific evidence to back it up.

clocks by? People living north of the Arctic Circle (or working on research stations south of the Antarctic Circle) experience perpetual dark and perpetual light at different times of year. This can mess with their body clocks, and delayed sleep or 'midwinter insomnia' is relatively common during the time of year with no natural light. Without the bright morning light, it's harder for the SCN to know how long to wait before initiating the sleep signal, and this can lead to problems in dropping off.

But despite this, most people still manage to follow a normal sleep routine. What is important when you can't rely on changes in sunlight to set your SCN is to give it other regular signals. This might be the times at which you eat, work or socialise. By keeping these constant you give your body the best chance of knowing when you should be sleeping, without the light cues it normally relies on.

Someone who is very familiar with this experience is physicist and broadcaster Helen Czerski, whose work has taken her north of the Arctic Circle a number of times. During her first experience of 24-hour sunlight she was camping on the side of a glacier, her only respite from the light a blackout sleep mask. More recently, she worked on board polar research vessel *Oden*, where more control of the light was possible. The lights would be dimmed in the evenings, she told me, and at night blackout blinds allowed those on board to sleep in total darkness. In both these cases, Czerski had no problem dropping off at night, and woke feeling rested each morning. She puts this down to the strict schedule she and the other scientists kept during both expeditions. 'Humans in those environments work so hard to maintain a 24-hour cycle. The first thing after feeding people is keeping a regular controlled schedule.' Czerski told me that all the scientists working on her expedition would keep to the same schedule of work, socialising and mealtimes. 'The way you keep everyone on the same schedule is you feed them at very strict times. Every ship I've ever been on has made it very clear: these are meal times. If you do not turn up at these meal times you do not get fed. You do everything you can to give your body a 24-hour cycle.'

This strict schedule gives their brains regular cues to keep them on the same 24-hour cycle, and she found it worked well, keeping her sleep and other daily rhythms on track: 'I certainly never felt hungry between dinnertime and breakfast. But then the schedule was so rigid – the trick is not giving your body a choice.'

Interestingly, Czerski told me that she still had a sense of time passing, even without the usual changes in daylight. In most places with 24-hour light, unless you are right on the North Pole, the sun does move across the sky, and its height also changes. On the glacier, she found herself able to use these cues to keep some sense of where she was in the day. But the boat brought another challenge, as it moved position every few days as well, meaning the path the sun traced across the sky would appear to change. Despite this, her brain still latched on to these cues: 'Even if the sun is moving to and from a different place each day because your ice floe has moved, you do have a sense of time passing.'

But it's not just the 24-hour cycle you need to mark when spending weeks drifting on arctic ice floes. Longer time periods can start to blur as well. On *Oden*, they were aware of this, and attempted to mitigate it by holding special dinners on a Saturday night. Rather than the usual canteen setting, the researchers were treated to tables laid with white tablecloths, and people made an effort to dress up and spend time over dinner chatting. 'Those excuses to mark the moment, to feel like time is passing... those are really important in long periods, where at least you know Saturday night has come round again.'

This sounds strangely familiar to me. As I write this, millions of people around the world are living in lockdown, to try to slow the spread of Covid-19. For me, as someone who has worked from home for a long time, my day-to-day life hasn't changed too much. But I have noticed the days beginning to blur into each other. To try to prevent this, my husband and I have started getting a takeaway on a Saturday night* to mark out the weekend as something different from our weekdays. This is helping to make it a little easier to keep track of the weeks in this strange time.

My own experiences with jet lag, and the science we have uncovered, make it clear that our body clocks are a vital element in how we know when to sleep. But why is this the case? Why not just base when we sleep on light levels, or simply on how long we have been awake? To understand this, we have to start with one rather surprising member of the animal kingdom: the Mexican albino cavefish.

Underground, overground... swimming free

Deep in the underground caves and caverns of Mexico there is a very special type of fish. It might not be the

* Though I must admit, we haven't quite stretched to the white tablecloth.

flashiest fish in the river, but *Astyanax mexicanus* is certainly one of the most interesting. This is because it comes in more than one type. These types are the same species, and could inter-breed if they wanted to, but they live very different lives. One inhabits rivers in Mexico and across the south of the US and has a typical fishy life, swimming around, looking for plants or small animals to munch on. But the other, which may live just a few metres away, has adapted to a very different lifestyle.

Unlike their surface-dwelling relatives, these cavefish live underground, in the many waterways running beneath Mexico's surface. It is thought they have lived these subterranean lives for half a million years, and so they have changed in appearance and behaviour compared to their aboveground neighbours. For one thing, they have lost their eyes. There is no need for eyes when you live in eternal darkness. They have also lost all the markings on their skin. Why bother with melanin to protect you from the sun if you live underground?*

These two groups of fish provide researchers with an unusual insight into evolution, genetics and, crucially to our story, the circadian rhythm. Normally, to work out what a gene does, scientists have to find or create a mutant animal with changes to that gene and compare them to the 'wild type' that doesn't have that change. So

---

* Interestingly, scientists used to think this was an example of a trait disappearing simply because it no longer provided a benefit, but recent evidence suggests there may be more to it. It seems melanin is made from the same precursors as a lot of brain chemicals, so if you stop making it, there is more of this precursor available. This could boost the amount of various neurotransmitters, which may provide a benefit to the animal.

if a normal fly has red eyes, but a fly with just one gene different has black eyes, you can say you have found a gene involved in eye colouring. But it's unusual to find two similar animals, with just a few genetic changes, living right next to each other. The cavefish provide a handy natural experiment.

The common hypothesis for why animals (and plants) have circadian rhythms is that it keeps us in sync with the cycle of night and day on this planet we call home. And this is supported by the fact that, as we have seen, we can use light to reset our natural clocks, ensuring we stay in tune with sunrise and sunset. So it might be easy to assume that a cavefish, living in the darkest depths, wouldn't need a circadian rhythm. In its rocky home, it is protected from cycles of light and dark, food is available throughout a 24-hour period, and there isn't even much of a temperature change to signal where the earth is on its daily rotation. Surely, in the same way it lost its unnecessary eyes and pigmentation, over thousands of years of evolution the cavefish would have lost its biological clock?

Andrew Beale investigated these fish for his PhD, taking samples from fish in the lab to test their gene activity (something which shows a circadian rhythm in other animals) at different times of day. And he discovered that these blind fish *do* have daily rhythms. Even more astonishingly, they could reset them with light, which they detect through photosensitive pigments in their skin. In their everyday lives, in the caves, this has been harder to test but findings suggest they *do* have a rhythm, though it may be weaker. But the bigger question was why. Why would something that lives underground, sheltered from the rhythms of the earth, have retained a circadian pattern?

Beale believes there is something more fundamental going on than we had ever realised. Perhaps rather than being an external force that we need to synchronise ourselves with, consistent sleep patterns allow our bodies to sync *internally*. As he explained to me, 'Every cell in the body has a rhythm, and even if they don't use the sun to synchronise these, it is vital for organisms to do so. So perhaps sleep arose in response to this grouping of processes, ensuring the body's cells all carry out their maintenance simultaneously, rather than in conflict with each other.'

These essential processes could be cell growth versus cell death, or at the level of the brain, taking in new information versus processing and storing memories. If the brain carried out both at the same time, it could lead to chaos, with fragments of your current surroundings being stored within memories from the day before. So perhaps the circadian rhythm evolved to ensure these vital processes take place at different times, so they can't interfere with each other.

It's not just these particular cavefish that have a body clock. Though not many studies have been carried out, it seems that other cave-dwelling animals retain a circadian rhythm of sorts too, although it is often weaker, and may not respond to light. And we now know that plants, and even single-celled organisms like some bacteria and yeast, have a version of a circadian rhythm as well. While we don't yet know for certain why, it definitely seems that having a pattern of some kind to follow is vital for life on earth.

But first, coffee

Despite their importance, our circadian rhythms aren't the only thing that controls how sleepy we feel, otherwise we would never nap after a late night, or find our eyes drooping during a boring meeting. There are other processes going on that interact with our body clock to tell us how long we have been awake and, surprise surprise, these too are controlled by chemicals.

It seems pretty obvious that the longer you have been awake, the sleepier you feel. This is known as 'sleep pressure' and is driven by a chemical called adenosine. This tiny molecule is a by-product of the metabolic processes our cells undergo all the time when we are awake, so it builds up throughout the day. The more of it there is sloshing around in an area of the brain called the basal forebrain, the higher our 'sleep pressure' and the more likely we are to drop off.

It is the adenosine system that so many of us trick with our daily dose of caffeine. When adenosine levels are low, acetylcholine is released in the basal forebrain. As we saw in Chapter 2, acetylcholine guides our attention towards external stimuli, making us feel awake and alert. Adenosine prevents the release of acetylcholine, so we feel sleepy. But caffeine blocks the receptors that are normally triggered by adenosine, making it seem to our brains that the concentration is lower than it actually is. This, in turn, means more acetylcholine is released, activating neurons in the cortex and making us feel more alert.*

But if you are one of those people who stumbles groggily out of bed in the morning, desperate for that first cup of coffee, beware – your body may have fought

* There may also be a complementary effect of caffeine on the circadian rhythm, as studies have found it can also delay the release of melatonin.

back! The levels of chemicals in our brains are carefully controlled, so if you regularly increase the level of one artificially, the brain often counteracts it by reducing the amount it makes, or changing how it responds to it. For example, while we aren't sure exactly how it does it, one way the brain might counteract caffeine's blocking of the adenosine receptors is by making more receptors.

This is known as tolerance, and is the same process seen in some people who regularly use illicit drugs (see Chapter 3). In humans, caffeine tolerance can build up in just a couple of days. Once you have become tolerant to caffeine, it no longer gives the same energy boost. In fact, you will feel more tired and sluggish before you have your coffee than if you had never drunk the stuff. All that first cup does is bring you back up to your baseline.*

But everyone knows that all the coffee in the world can't keep you awake for ever, so there must be more going on than just this one chemical. And the build-up of adenosine doesn't tell us *how* we fall asleep. That process of transitioning from consciousness to unconsciousness is incredibly complex, but we are starting to understand it thanks, in part, to a First World War fighter pilot, doctor and neuroscientist who died nearly 100 years ago.

The sleep switch

In 1916, a strange disease began to present itself in Vienna. Patients suffered from a short, acute illness,

* You can reset your caffeine tolerance by giving the stuff up. Unfortunately, it can take a couple of weeks of abstinence for your brain chemistry to reset itself, and during that period you might suffer withdrawal effects, ranging from irritability to headaches.

feeling generally unwell with a headache and a mild fever. But then they would start to feel sleepy and confused, beginning to spend less and less time awake. Many also experienced tremors and weakness of the hands or eye muscles. After a couple of weeks of these symptoms, around half of people would recover while others would fall further and further into sleep. Where at first they had been easy to rouse, over days or weeks they would become more delirious during their short periods of wakefulness, before slipping into a coma. Death would often follow.

At this time, a young man was working as a pilot on the front line for the Austrian air force. His name was Constantin Von Economo, and he was the youngest child of Greek aristocrats, though he had been born in Romania, grown up in Italy, and then moved to Vienna to study. It was these studies that brought him home from the front line – along with the worries of his parents. Von Economo was a doctor, and so, pressured by his parents to stay out of danger, he returned to Vienna to work as a military physician caring for people with head injuries. It was here that he first saw patients with this mysterious illness. Realising this was something mostly unknown to science, he wrote a paper describing the disease in 1917 and gave it the name 'encephalitis lethargica', although it was more commonly known as 'sleepy sickness'.*

* This is very different from sleeping sickness, a disease caused by a parasite and spread by tsetse flies, which causes symptoms starting with fevers and headaches. If left untreated, the parasite can cross the blood–brain barrier, leading to confusion and the disrupted sleep patterns that give the disease its name.

As well as the drowsy type, he recorded two other types of the illness in the paper, each with its own constellation of symptoms. The second form made patients restless, with physical twitches and mental frenzies, along with severe pain and, often, insomnia or a reversal of the normal sleep pattern. In some cases, this form later turned into the sleepy form, but sudden death could also occur at any stage. The symptoms of the third type were similar to those of severe Parkinson's disease, with weakness and rigidity in the muscles a prominent symptom. Although less likely to be fatal, this version often became chronic, with patients unable to return to their pre-illness lives.

Von Economo was fascinated, remembering a similar case from his childhood, which had occurred after a flu pandemic. And so he assumed, as many have since, that this outbreak of encephalitis lethargica, the worst the world had ever seen, was due to the Spanish flu which, at this time, was killing more than 50 million people worldwide. He wanted to know what was causing this range of symptoms, so he began analysing the brains of people who had died of the disease.

He realised that these patients, with their severe sleep disturbances, could tell us something about how sleep is controlled in the brain. He used this 'natural experiment' to identify an area of the brain that controls the switch between sleep and wakefulness. In doing this, he was the first to explore the importance of regions we now call the hypothalamus and brain stem. He noted that his lethargic patients had damage to the back part of their hypothalamus, while in people with insomnia the damage occurred further forward in the hypothalamus, near the optic nerve. He argued that these two areas were the 'sleep part' and the 'waking part' of the brain.

It is thought the illness killed a million people in around 10 years, but the cause of 'sleepy sickness' is still somewhat of a mystery. More recent analysis hasn't supported a direct link to the flu, but no other definitive cause has been found. Some researchers argue that it is an autoimmune disease, where the body itself attacked and started breaking down brain cells, perhaps triggered in some by the flu infection. Others believe an as-yet-unidentified virus was to blame. Fortunately, the disease has all but disappeared, but this means there is very little chance of solving the mystery, unless sleepy sickness rears its ugly head again in the future.

He may not have discovered why people developed the illness, but when it comes to his findings about sleep, Von Economo was right on the mark. In 1929, after the disease had vanished as quickly as it had arrived, he presented his deductions to the College of Physicians and Surgeons in New York. He argued that these two regions of the hypothalamus were the brain's sleep and wake centres,* and that the neurons in these centres would initiate a chain of events to damp down or ramp up activity in the cortex, allowing us to transition from wake to sleep and back again.

Amazingly, in the nearly 100 years since he put forward this theory, a wealth of evidence has accumulated to back it up. Animal studies have found that damage to the front

* Although not common in general speech, scientists refer to the state of being awake as 'wake', in contrast to sleep, so I will use these terms throughout this chapter.

part of the hypothalamus induces insomnia, while damage further back will cause an animal to sleep much more than is typical. And if electrical stimulation is applied to the hypothalamus, an animal can be made to fall asleep or wake up, depending on where that stimulation is applied.* Modern imaging studies in humans have also confirmed these areas are vital for sleep and wake.

Sleeping peacefully?

The next question, of course, is why? What is it about these regions that makes them the sleep and wake centres of the brain? We now know that the hypothalamus contains neurons which use both sleep- and wake-promoting chemicals, and this is why it is of such importance.

The systems that control our transition from sleep to wake and back again are incredibly complex, and we don't yet fully understand their intricacies, but we have a general idea of the ways in which they work, and the chemicals that are vital for them to do their job. You can think of sleep and wake as a see-saw, two states that can be flipped between by the smallest of changes.

---

* Whether this works in humans isn't clear, as deep brain stimulation is reserved for severe brain diseases, because of the risks involved. One of the diseases it is used for, however, is Parkinson's, and the electrode is often implanted in the subthalamic nucleus, not far from the hypothalamus. As well as the improvement in motor symptoms that the procedure was designed to provide, many patients experience an improvement in sleep quality too. However, we can't know if that is a direct result of the stimulation, and/or thanks to a reduction in other symptoms, and/or the medication previously needed to control them. A much safer option is providing electricity through the skull, rather than deep inside the brain, and studies are just beginning to look at whether this might help treat insomnia.

On one side of the see-saw is the arousal system. This is a collection of neural circuits, beginning in the brainstem and with neurons that reach throughout the cortex. Neurons in these circuits use a wide range of neurotransmitters: dopamine, noradrenaline, serotonin, glutamate, histamine and, perhaps most importantly, acetylcholine. As we saw earlier, acetylcholine activates neurons in the cortex, helping us feel awake and alert. All of these complex pathways pass through the hypothalamus at some point on their journey, right through the area von Economo called the 'wake centre'. This is also where melatonin seems to come in. Though research on its mechanisms is only just beginning, some studies have found that melatonin appears to block some of the neurons in the 'wake centre', helping you slip into sleep more easily.

On the other side is a system of neurons which use the chemical GABA. GABA inhibits neurons in the cortex, making it harder for them to reach the point of firing. This damps down activity in this area, triggering sleep. These neurons originate in a part of the hypothalamus which sits, you've guessed it, right in von Economo's sleep centre.*

When we fall asleep or wake up, our brain must flip the 'switch', and tilt the see-saw. To prevent this happening too readily, the two systems inhibit each other: when the arousal system is active, serotonin and noradrenaline dampen down the activity of the 'sleep centre', and when the sleep circuit is active, GABA reduces activity in the 'wake centre'. This means that

---

* Although, as ever, things are a little more complicated here, as while GABA in the hypothalamus promotes sleep, GABA release in an area of the brainstem can encourage wakefulness.

(hopefully) you don't fall asleep too often when you should be awake, and you aren't constantly woken up when you should be sleeping. It seems likely that how easy it is to flip this switch can vary between people, depending on differences in these brain circuits, or levels of the chemicals involved. So maybe this is where my issues lie – my switch is harder to flip, my see-saw biased towards wake rather than sleep.

But this isn't quite as simple as the see-saw you might see in a child's playground. As well as these two competing systems, there is a third, stabilising system, consisting of neurons in the hypothalamus that use a neurotransmitter called orexin.* These neurons connect with the wake circuits in the cortex and throughout the brain, helping to boost these areas during wake. They receive input from areas controlling circadian rhythm as well as information about metabolic processes in the body, so they may be one of the ways these factors influence sleep. When we sleep, these neurons are inhibited, but when we are awake they are active, keeping us alert.

Patients with narcolepsy, who often slip into sleep many times throughout the day, with no control over the process, have been found to have reduced numbers of these orexin-producing neurons. This suggests they have a role in preventing the sleep switch from being triggered too easily.

---

* The same chemical is also called hypocretin as it was discovered and named simultaneously by two different research groups. The two are used interchangeably, but I will stick with orexin for simplicity's sake.

This sleep switch and its stabilisers drive the brain changes that tip us from wake to sleep and keep us there (mostly) throughout the night. But even when you are asleep, the job of your brain chemicals is far from finished. In fact, levels are constantly changing throughout the night.

When we look at human sleep over the course of the night, we see a distinct pattern. Sleep isn't a simple on/off process; instead, you pass through various stages, and each of these is characterised by a different pattern of brain activity, and a different combination of brain chemicals. At the start of the night, you transition through light sleep into deeper, slow-wave sleep. During this phase, GABA levels in the cortex are high, reducing the excitability of our neurons. Your brain is at its least active, with neurons firing in rhythmic patterns, producing the slow waves of electrical activity that give this phase its name.

But this stage doesn't last for long. After an hour or so, you will enter the next major phase of sleep: REM. REM stands for rapid eye movement, named because in this type of sleep your eyes dart around under your eyelids. Your brain is also much more active. EEG recordings of the electrical activity in REM sleep look very much like the waking brain. Many of the patterns of neurotransmitter activity are similar too. In both REM sleep and wake, GABA levels in the cortex are low, meaning the neurons can be highly active. Acetylcholine levels are high, as excitatory synapses in these neurons talk to each other, activating the cortex. Experiments have found that a drug that increases the amount of acetylcholine at the synapse can tip humans from slow wave to REM sleep.

The difference between wake and REM sleep becomes clear, however, when we look at levels of the monoamines, including serotonin, noradrenaline and histamine. When we are awake, levels of these chemicals are much higher than when we sleep, all across the brain. This cocktail of brain chemicals keeps us bright-eyed and bushy-tailed, preventing us from falling asleep when we don't want to.

Histamine in particular is released in greater amounts when you are actively vigilant, compared to quiet wakefulness. This may sound surprising – histamine is better known as the chemical that makes us itchy when we are allergic to something. Blocking its effects, using an antihistamine drug, can help relieve symptoms of hay fever, pet allergy or whatever else is triggering your immune system to overreact. But in the brain it has a different role to play, encouraging alertness. This is why many of the older antihistamine drugs, which could cross the blood–brain barrier and block histamine receptors in the brain, made people sleepy. Modern allergy drugs tend to contain antihistamines that can't make it into the brain, so don't have this effect. But the old ones are still in use; they are the main component of many over-the-counter sleep aids, because of histamine's role in preventing us from entering the Land of Nod.

The night time is the right time

Now that we have a better grasp of the way our brain chemicals and systems control sleep, this leads us to an even more fundamental question. Why do we need to sleep? As anyone who has ever written a book while also trying to continue a career in presenting/school shows/training/a million other things can attest, sleep can feel like a hindrance. I have often found myself wondering, staring at a to-do list as long as my arm, what life would be like if I didn't need to sleep. Sadly, we know that adult humans do need their eight(ish) hours if they want to function properly. I know for sure that I can't do anything that requires my brain after a poor night's sleep. While I may be able to get through the day, zombie-like, after a long-haul flight, the idea of writing or researching complicated topics like sleep is completely out of the question. So, begrudgingly, I do my best to ensure I get the sleep my brain needs in order to work at its optimal level.

But why? Why would we have evolved to need a period of the day when we are unable to hunt, eat, have sex or do any of the other things that would have helped our ancestors to survive and pass on their DNA? Even worse, during this time we are (largely) unaware of what is going on around us and unable to move, surely leaving us more vulnerable to predators. For sleep to still be advantageous despite these issues, there must have been a serious benefit. So what is it? This is one area where scientists are still scratching their heads. But there are lots of hypotheses.

Perhaps the simplest idea is that we sleep to conserve energy. To understand this idea fully we need to travel back to the time of our ancient ancestors, who lived in small groups, hunting and foraging to survive. Humans are incredibly reliant on sight, so we are unlikely to be very successful trying to find food at night. And if you aren't going to be able to find food, you may as well stay very still, so that you burn as few calories as possible. Settling down for the night somewhere safe and hidden, such as in a cave, would also have the advantage of keeping you out of harm's way; you are less likely to be eaten by a nocturnal predator if you keep hidden than if you are wandering about randomly in the dark! The problem with this theory, however, is that it doesn't explain the loss of consciousness that comes with sleep; surely being still, hidden but fully aware of your surroundings, would be the safest course of action of all?

You may think that, with our brain being a power-hungry organ, using up about 20 per cent of the calories we consume every day, switching it off would save energy, like turning your phone off when it is low on battery. The problem is, this isn't the case. As we have seen, our brains are highly active while we sleep, and sleeping for eight hours only saves about 135 calories over sitting quietly awake. That's the equivalent of one and a third large apples, or one bag of ready salted crisps. I don't know about you, but to me this doesn't seem like a good enough reason to take the risk of becoming unconscious for large portions of the night.

From an evolutionary point of view, if we have evolved to do something risky, it must be because the benefits outweigh the risks. The classic example of this is sex. In many animals, sex is a seriously risky business. Males may have to face off competitors in fights to the death, or it may be the object of his affections herself who poses the threat (as in praying mantis and many species of spider, where the female will often treat the male as a tasty snack, either before, during or after sex). Females take a risk too – developing eggs or growing a baby animal inside you takes a lot of resources, and for humans in particular, childbirth has, historically, been extremely dangerous. Even now, with all the advances in modern medicine bringing mortality rates down dramatically, it still poses over 10 times the risk of skydiving.* So why do we do it? Because the evolutionary benefits outweigh the risks. If we get to pass our genes on to the next generation, whatever it takes was worth it.

It's not just activities that directly help an animal to pass on its genes that have been selected for by evolution. In order to procreate, you need to live long enough to meet and seduce a partner (for most animals at least!). So taking risks can also be worthwhile if they help you *survive*. That's why animals take risks in search of food or shelter. As sleep doesn't seem to directly help the sex life,† perhaps there is something that occurs during the process of sleeping that is needed for our survival. This is the claim of the next set of theories we will look at.

Beauty sleep for the brain

To understand the role of sleep, scientists have begun looking at what is actually going on inside the sleeping brain, trying to discover some function that must be carried out daily to maintain our brains. As we have seen, the brain is far from restful during the wee small hours. So what's going on in there?

One theory is that sleep evolved because of its importance for learning and memory. If you aren't getting enough sleep, you will find your memory suffers. This is

---

* David Speigelhalter's book, *The Norm Chronicles*, puts giving birth (vaginally) at 120 'micromorts' while skydiving has a value of 10.

† In fact, if your partner has ever rebuffed your amorous advances, claiming they are 'too tired', you might be of the view that it actually hampers them.

something that is experienced by sleep-deprived new parents everywhere, who blame 'baby brain' for forgetting where they left their keys. Studies have found that if you are sleep deprived, not only does your concentration suffer, meaning you can't take in new information as efficiently, but connections in your brain also form less well, particularly in the hippocampus. As we saw in Chapter 2, these connections are vital for new information to be stored, so a sleep-deprived brain is not only badly placed to take in new information, it is also unlikely to be able to store it. It seems getting enough sleep is vital for us to be able to learn and remember.

And it's not only sleep *before* the learning that is important; sleep has another role to play *after* you have learned something new, in helping ensure you retain that information.

In fact, if you take a nap after learning something, you are likely to remember it better than if you stayed awake. Scientists used to believe this was simply because when you are asleep you aren't taking in any new information, so there is nothing to interfere with whatever it was you just learnt. But we now know there is a more active process going on.

While you are awake, the brain is primed for taking in information, but it is while you sleep that it can organise and store that information. During the night, your hippocampus, buried deep within your brain, repeatedly activates areas of the cortex, transferring memories from temporary to stable, long-term storage (see Chapter 2). It is also during sleep that the memory is processed, the gist is extracted and it is fitted in with other information. This is why you can often experience a moment of insight after 'sleeping on' a problem. Some people even believe that dreams are part of this process, with the brain replaying information from the day in various ways as it stores it.

Another clue lies in the way our sleep changes throughout our lives. Babies sleep a huge amount, possibly because they are learning so much when they are awake. Young children also need plenty of sleep, and even teenagers need more than adults. Could this be because our adult brains have less memory processing to do during the night? We don't know for sure, but it is an interesting possibility.

As we age, our sleep changes again. Older people tend to sleep less, and the sleep they do get is more fragmented. There is a common misconception that this is because older adults need less sleep, which would fit with the idea that they are taking in less new information, so don't need as much time to store it. Recent studies, however, have shown that this isn't true. For example, a study by Iris Haimov, at the Max Stern Yezreel Valley College, Israel, found that older adults who have more disturbed sleep have more memory problems. Insomnia has even been linked to a greater risk of death, even when age and existing health conditions are taken into account. It seems older adults need just as much sleep as younger adults, they just struggle to get it.

In part, this is down to changes in our body clocks, which shift to an earlier release of melatonin as we get older. This makes many older adults sleepy in the early evening, but if they push through (perhaps with the help of a nap, or dozing in front of the TV) this can lead to trouble falling asleep when they do go to bed. Despite this, they may wake early, not spending enough total time asleep. There are other problems too. As we age, our bladder becomes weaker, so we are more likely to wake throughout the night to use the bathroom. The body clock's signals also become dulled, not having as much of an effect on the brain's desire to sleep as they did when the individual was younger.

Older adults tend to spend less time in deep, slow-wave sleep, and more time in lighter sleep, from which they are more likely to be woken by noises or other disturbances around them. This may be down to changes in the brain's sleep and wake circuits. As we age, we begin to lose neurons, and there are areas of the sleep and wake circuits where this seems to be particularly common, including the orexin-producing neurons in the hypothalamus. This loss might destabilise the see-saw, making it easier for older people to be flipped from sleep to wake and back again. As we age, the number of adenosine receptors also decreases. This means that although the chemical builds up as normal (or even more so) throughout the day, older adults don't feel as much of a difference in sleepiness depending on how long they have been awake. And, unable to respond to the adenosine, their brains don't sleep as long or as deeply as they should to clear it.

As we have seen, we need to be well rested to learn, and sleep after learning is vital to store new memories. So there is an interesting possibility here: could the cognitive decline that often comes with age be *caused* by this fragmented sleep? It's hard to know for sure, as it could simply be a correlation, or the link could be the other way around, with cognitive issues causing the sleep disruption. To work out which was the case, we would need to do an experiment, taking healthy people, and waking them throughout the night, for years. I'm sure you can see the problem here – can you imagine being asked to take part?

> 'We would like to limit your sleep to only four hours a night, by waking you at regular intervals, and see how that affects your brain, and whether you are more likely to develop memory problems.'
>
> 'Hmm – not sure about that, but I guess it is for a good cause. How long will the study go on? Two days? A week?'
>
> 'Fifty years…'

The brain's rinse cycle

Instead, scientists have begun looking into what is going on in the brain when animals sleep. In 2013, a paper was published that changed the way we look at sleep and the brain. Maiken Nedergaard and colleagues at the University of Rochester, US, were looking at the flow of cerebrospinal fluid. This clear liquid surrounds your brain, cushioning it against impact and filling the gaps (or ventricles) within it. The team noticed that when mice were awake, the fluid hardly moved throughout the brain, but when they slept, it started circulating much more. They discovered that brain cells shrank when the mouse was sleeping, widening the spaces between the cells, and allowing the fluid through.

Although it sounds negative, this shrinkage is actually a vital process for keeping the brain healthy. During the day, metabolic processes lead to the build-up of all sorts of waste products in the brain. This night-time 'rinse cycle' washes away the waste, allowing the brain to function efficiently the next day. The researchers liken this to a dishwasher turning on, and named the system 'glymphatic', based on the body's similar lymphatic system* and the glial cells in the brain which seem to control the process.

Of all the products rinsed away by the glymphatic system, the one that sparked the most interest is a small fragment of protein known as amyloid-beta (A-beta). The reason for the interest is that A-beta is thought to be associated with Alzheimer's disease, particularly when it clumps together into large 'plaques'. While we don't know the exact mechanism behind the link, the brains of people with Alzheimer's are often found to have more of these A-beta plaques than the brains of healthy people of a similar age. This raised a fascinating possibility. If sleep is a time when the brain washes away these potentially toxic molecules, could sleep-deprivation put an individual at risk of developing dementia? Could this be another way in which lack of sleep can lead to cognitive decline?

There are plenty of studies that find a link between a lack of sleep and the build-up of A-beta plaques in the brain. When healthy older people are followed over years, those who report worse sleep and more daytime sleepiness showed more A-beta plaques in neuroimaging studies and more Alzheimer's bio-markers in their spinal fluid. But the case is far from clear cut. Alzheimer's disease itself causes sleep problems, and those people who felt sleepy during the day tended to have more

* The lymphatic system is involved in maintaining fluid balance and is an important part of the immune system. It also helps the body to get rid of by-products, damaged cells and bacteria or other pathogens.

plaques in their brains to start with. So, as we saw earlier, it is difficult to know whether the sleep problems are a cause or an effect of the brain changes seen. And these changes build up over decades, so again, it's difficult to devise an experiment to test it directly.

Because of this, scientists are stuck trying to deduce what's happening from the information we can get. And one thing that it is possible to do is to see what happens in the brain after a single night of sleep deprivation. Astonishingly, a small study, led by the director of the US National Institute on Drug Abuse, Nora Volkow, found a 5 per cent increase in A-beta in the hippocampus (which as we have seen is important for memory) after just one night of sleep deprivation. Another way to overcome this problem is to look for natural experiments – individuals who may have been suffering sleep problems for years, who can be compared to similar individuals without the problems. One group of potential interest is people who experience sleep apnoea.

Sleep apnoea is a relatively common condition where the throat closes over during sleep, causing sufferers to wake, gasping for breath, before falling asleep again. This can happen many times a night (sometimes as often as every two minutes), and the person usually doesn't remember it happening in the morning. They do, however, notice that their sleep is unrefreshing, often waking up feeling like they haven't slept a wink, with no idea why. There is a strong link between sleep apnoea and mild cognitive impairment (which is often a precursor to Alzheimer's), and there are even hints that treating the breathing problems and so restoring restful sleep might help improve cognitive symptoms.

Of course, we can't rule out that it is the lack of oxygen to the brain, rather than the lack of sleep, that causes the problems, but it is certainly an interesting avenue of research, which might help us better understand why sleep is so vital for our brain's health.

Sleep to forget, sleep to remember

As well as its role in memory and cognition, sleep is also vital to keep us emotionally stable. We all know that overtired children* can be whiny, tearful or irritable, but the same goes for adults. Research at the University of California, Berkeley found that just one night of sleep deprivation increased anxiety in subjects, with many even reaching the clinical criteria for an anxiety disorder. When the scientists looked at their brains, they found increased activity in the amygdala and other emotional brain regions, and reduced activity in the prefrontal areas that connect to the amygdala and regulate our emotions. As we saw in Chapter 4, it is the balance between these two areas that helps us control our emotions, so if the prefrontal areas are no longer able to inhibit the amygdala's reaction, this could explain the mood swings that are so common in sleep-deprived people. It also raises interesting questions about mood disorders, and whether sleep problems could contribute to anxiety, depression and other disorders, where there is an imbalance between these two systems.

There is also another effect of sleep deprivation on mood. Not only does it make you more reactive, but losing a night's sleep seems to bias you towards seeing things as negative. In one study, led by Daniela Tempesta at

* Or husbands, in my experience!

the University of L'Aquila, Italy, researchers took healthy students and divided them into two groups. One group was allowed a normal night of sleep, while the others stayed awake. Both groups were then shown six video clips, two positive, two negative and two emotionally neutral. Later, after both groups had slept normally for two nights, they came back to the lab and were tested on how well they remembered the clips. Unsurprisingly, the initially sleep-deprived group did worse on the memory test overall. But the difference disappeared when the researchers looked at just the negative videos. It seems that while losing sleep will make it harder for you to remember good or neutral things, anything bad that happens to you will still be stored. Or, as the researchers put it in their paper, 'The recognition of negative stimuli is more "resistant" to the disruptive effects of sleep deprivation.'

This makes a lot of sense. We know that sleep deprivation leads to overactive amygdala responses, and we know the amygdala helps us to efficiently store emotional memories. It also makes sense evolutionarily. It is a lot more dangerous to forget that you were nearly killed by a sabre-toothed cat (negative) when you went to a certain place than to forget you saw a bird (neutral) or found some tasty berries (positive).

Tempesta and her colleagues have also found that when sleep deprived, people have a more negative outlook than when they have slept well. And that sleep deprivation doesn't have to be a completely missed night. In a recent study, people rated positive and neutral images as more negative when they had had their sleep restricted to five hours a night for five nights than when they had slept normally. And a study led by Jaime Tartar at Nova

Southeastern University, US, found that students with poor sleep quality were more likely to have depressive symptoms, as well as showing this negative bias.

If this all sounds familiar, it should. This negative bias is extremely similar to that found by Catherine Harmer in people with depression (see Chapter 4). And this isn't particularly surprising, as there are strong links between sleep problems and almost every mood disorder in the book. It might even be a vicious cycle. Not getting enough sleep could lead to this negative bias, making someone feel their life is worse than it actually is, increasing their anxiety or depression, which in turn makes it harder to sleep.

Many antidepressants seem to improve sleep quality, and this might be one mechanism by which they work. In fact, it does seem that those whose sleep normalises while on antidepressants are more likely to recover and less likely to relapse.* Ironically, though, this doesn't happen in everyone, and insomnia can also be a side effect of antidepressant drugs in some cases, showing the complexity of these systems.

Another idea suggests that while we sleep, memories are stored, but their emotional content is reduced. Matthew Walker, Professor of Neuroscience and Psychology at the University of California, Berkeley and author of *Why We Sleep* put forward a hypothesis he terms 'sleep to forget and sleep to remember'. When memories are first formed, they are rich with emotion, whether that is positive or negative. But, over time, the emotion fades, so while I know that my first kiss was thrilling and terrifying in equal measures, I

---

* Although, again, it is hard to know the direction of the relationship. It is possible that a reduction in depressive symptoms made it easier for the individuals to sleep, rather than the other way around.

don't re-experience that emotion today in the same way that I did in the days following it.

Walker argues that while we sleep the emotion of a memory is 'decoupled' from its content, so you can remember what happened without experiencing its emotional impact. If this process went wrong, you could end up being flooded with emotions every time you recall a memory, like the people with PTSD we discussed in Chapter 2. In particular, he argues this decoupling process happens during REM sleep. This would be a particularly good time for it to happen because of the neurotransmitters active in the brain. During REM, just as when you are awake, GABA in the cortex is low and acetylcholine is high, so the neurons can be reactivated in a similar way to when the memory was formed. But levels of noradrenaline, a stress hormone, are much lower than in wake, as is activity in neurons that use other monoamines like dopamine and serotonin. This offers the opportunity to replay a memory without the emotional experience. Despite the attractiveness of this hypothesis, and Walker's claims in his book, whether this actually happens is still under debate. The results of experimental studies are mixed, with some finding either sleep in general, or specifically REM sleep actually *increases* the emotional content of memories, rather than reducing it.

Even less well understood is whether dreams are the manifestation of this emotional processing. Certainly, a large proportion of dreams have emotional content, and you are more likely to dream about something if you are worried about it. There is also some evidence that people who dream about difficult issues more often recover better from the emotional conflict or trauma. Emotional dreams could give us the opportunity to practise our emotional regulation skills, meaning we are better equipped to deal with real stressors when they arise during the day. On the other hand, they could simply be a screensaver – something created by the brain to keep our conscious minds busy while they aren't receiving input from the outside world. Despite being a topic that has fascinated humans for thousands of years, we still know very little about the reasons behind our dreams and whether their content has any meaning at all.

Sleep – what is it good for?

While each of these theories provides an interesting insight into the processes that happen while we sleep, we can't know for sure which was the reason for its evolution. Part of the problem is that sleep has more than one function. If an animal spends part of its time motionless, and less responsive to the outside world, it would be advantageous to carry out all sorts of brain and body maintenance during this time, rather than when the animal was awake and busy hunting or mating. So once sleep evolved, it seems likely evolution would drive all sorts of other processes to happen during this time. But that means it is almost impossible to know which came first – which was the driving force for its evolution.*

* This is a real problem for evolutionary psychology in general. Unlike bones and teeth, brains don't fossilise, and it is hard to recreate an early human's behaviour from archaeological sites. This makes it difficult to ever prove or disprove evolutionary theories. While I have included some in this book, as I think they provide an interesting way of thinking about the brain, they should be taken with a note of caution for this reason.

Another problem with all these theories of sleep is that they focus on humans, and things that are important to us: emotions, memory, cognition. But this doesn't explain why sleep is so universal across the animal kingdom. And it can't explain the variation that is found between animals. If it were the case that we sleep for our brain to process information and store memories, or to deal with emotions, we would expect that those animals with larger and more complex brains, and higher intelligence, would sleep more than simpler animals. But that isn't the case. Koalas, for example, aren't the sharpest tool in the shed, yet sleep up to 22 hours a day.* Elephants, on the other hand, despite their intelligence and complex social lives, sleep for just two hours a day in the wild. In fact, there doesn't seem to be any correlation between sleep time and cognitive ability in animals, and sleep-like states have been observed in jellyfish, which don't even have a brain! These findings imply that though we may *use* sleep to process and store memories, and it may be important for cognition and emotion, this can't be the whole story. It must have evolved an extremely long time ago in our evolutionary history, and have been conserved, in some form, in every animal that has evolved since. This suggests the process that drove it must be vitally important, but for now, exactly what that was remains a mystery.

I had hoped that by this point we would have a better understanding of the processes that allow us to sleep, and what happens in the brain while we do. And that perhaps

---

* Although one study suggested some of this is just rest, and they actually only sleep for 14 hours a day. Though I'd say that's still an impressive amount!

I could use some of this knowledge to improve my own sleep. But while we are starting to understand how our brains move from a state of wake to a state of sleep, this hasn't really led to any revolutions in improving individual sleep quality.

Supplementing with melatonin might benefit older people, whose circadian rhythms are blunted, as well as people with disorders of their body clock. And there are various types of sleeping pills, which can be helpful for people with clinical insomnia. But these are usually only recommended as a short-term solution, to be used for the odd bad night, not every day.

Part of the problem is that without an understanding of what sleep is really for, it is hard to know how we can encourage the right type of sleep. Some drugs, for example, help people sleep, but change the architecture of that sleep, so they might have more REM than usual, or less. We aren't yet sure what the knock-on implications of that might be. Scientists are looking into other ways of encouraging the right types of sleep at the right times, including using electrical stimulation through the skull, or providing certain sounds at certain times of night, but these types of intervention aren't yet ready for general use.

So will I be changing anything to help improve my sleep? It seems that the best techniques are behavioural ones. Focusing on getting light in the morning to set my circadian rhythm for the day, not eating too late in the evening, and taking some time to wind down before bed, away from light-emitting screens, might not cure insomnia in all cases, but for many of us it just could be the difference between a good and a bad night's sleep.

CHAPTER SIX

Food for Thought

We've all been there. Midday is approaching, and it feels like breakfast was days ago. You are sitting in a quiet office, trying to focus on the meeting, but you can't help but be distracted by the growing sensations in your stomach. That queasy, gnawing feeling that you know means only one thing. Suddenly… grrRRrr… your stomach lets out a rumble* loud enough to alert everyone in the room to your hunger. Hunger is a vital drive – it ensures we eat enough to survive and reproduce. But it isn't as simple as it might first seem, and a lot of it is controlled by chemicals in your body and, perhaps surprisingly, your brain.

Hunger can be triggered by internal states like low blood sugar or a lack of fat reserves, but it can also be initiated by the smell or sight of food, or even by stress. But for me, it was something I didn't experience until I was 11. I always liked my food, and as a child I ate pretty much whatever was given to me. But unlike some of my friends, whose parents had to have bags of snacks on them at all times, I never demanded food. I remember, vividly, waiting in my year six classroom to be dismissed for lunch, turning to my friend and telling

---

* The technical term for a rumbling stomach, in case you were wondering, is borborygmus.

her that I didn't understand why I kept feeling weird before lunch – kind of sick, or like I had a tummy ache. She looked at me incredulously, replying, as if talking to a particularly stupid toddler: 'You're hungry.'

Since then, hunger has been a regular companion. I'm a grazer, and struggle to go more than a few hours without a snack, but get full quickly at mealtimes. In contrast, I have friends and family members who can skip lunch in order to really enjoy an evening meal out, and put away much bigger quantities, while I struggle to finish my main course. So why the differences? What is it, physiologically, that makes me unable to function without regular sustenance, while others don't feel hungry until they sit down to eat?

The need to eat

Before we can dive into hunger in the brain, we need to understand what happens in our bodies when we eat. Keeping the level of sugar in the blood stable is vital for the function of our body and particularly our brain. If it gets too high, for too long, it can damage the eyes, kidneys and nerves. And if it is too low, symptoms such as dizziness, blurred vision and even seizures can occur. Luckily, in most people, our bodies work to prevent either of these extremes from happening, without us having to think about it.

When we eat something containing sugar or carbohydrates, these molecules are rapidly broken down into glucose, which is absorbed through the lining of our small intestine into the bloodstream, causing an increase in blood sugar. Our pancreas detects the increase and releases a hormone called insulin, which encourages muscle and fat cells to take in the glucose. It also causes the liver to convert glucose into glycogen, which can be stored for later. But only a certain amount of glycogen can be stored; if we take in more sugar than the liver can handle, the remaining amount is converted into fat.

When blood sugar levels drop, so do insulin levels, and the liver begins to convert glycogen back to glucose, releasing it into our bloodstream. People with diabetes either don't produce insulin (type 1) or don't respond to it (type 2), meaning this process can go awry. The difficulty they have maintaining blood sugar levels highlights the importance of this mechanism in our body's health.

Insulin also has another role, in the brain. When levels in the blood are high, insulin travels to the brain, telling it you are full. Insulin's role here is vital, and mice with no insulin receptors rapidly become obese. As insulin levels fall, with falling blood sugar, our brain detects this and initiates a hunger response.

But this isn't the only signal our brain receives. In fact, whether we feel hunger or not is a delicate balancing act between a whole host of different signals. A few hours after you last ate, food passes out of your stomach and into your intestines for further processing. This leaves your stomach empty, and it shrinks. Sensors lining the stomach wall measure how stretched your stomach is and, detecting this change, send electrical signals directly to the brain, via the vagus nerve. Simultaneously, a hormone called ghrelin is released.

Ghrelin is the only 'hunger hormone' we have found so far, but it has a dramatic effect. If people are given extra ghrelin, they will eat around 30 per cent more. To see how powerful this hormone is, we can turn to people who produce too much of it. One group is those with a genetic disorder called Prader-Willi syndrome. People with the syndrome have learning disabilities, paired with severe obesity and voracious appetites. We now know that they have high levels of ghrelin, which may explain their overeating and weight gain.

So an empty stomach signals that we need to eat, but that's not the end of the story. As our last meal reaches the intestines, another signal is released. A chemical called CCK (cholesystokinin) triggers changes that help your body process and absorb nutrients from the food, but it also sends signals to the brain via the vagus nerve, which connects the brain with the digestive system. This tips the balance back towards feeling full rather than hungry. And it delays the emptying of the stomach, so the stretch receptors don't send their 'eat more' signals. The more fat your meal contained, the more CCK is released, and the less likely you are to eat again immediately.* After a big meal, CCK and other satiety signals can override those produced by the empty stomach, meaning you don't feel the need to eat again for several hours.

This multitude of signals, some carried in the blood and others via the nervous system, travel to the brain, where they must be interpreted – and it is the brain that drives hunger, as we feel it. So how does it decide which signal is most important? This, of course, is where things

* There are other chemicals involved in the sensation of fullness as well. The lower intestines release peptide YY, slowing down digestion, and another hormone called amylin is released from the pancreas. Both act on parts of the brain, helping us feel satisfied after a meal.

get a bit more complicated… but before we tackle that, time for a snack.

Tipping the scales

The hypothalamus is probably the most important brain region for controlling our appetite, and it is well positioned to be an 'eating centre'. It has connections with various other brain areas, and with the pituitary gland, meaning it can control the release of hormones. Unusually for a brain region, part of the hypothalamus is outside the blood–brain barrier, meaning it has access to your main blood supply, and can detect the levels of circulating glucose, for example. It also monitors the workings of your immune system. If there are high levels of pro-inflammatory cytokines in the blood, the hypothalamus knows you have an infection, and raises your body temperature to try to destroy it – you have a fever. It also reduces your appetite, so you can mobilise all your energy into fighting the illness rather than digesting food.

It was known as early as the 1940s that damage to this area of the brain could cause obesity, both in animals and humans. In experimental studies, animals which had these areas destroyed overate, seeming insatiable in their desire for food. But these techniques were crude, affecting the hypothalamus and the areas around it, making it difficult to know for certain which areas were involved. As experimental techniques developed, studies with more precise lesions showed that it was only part of the hypothalamus that, when damaged, produced overeating and obesity: the 'ventromedial' area.

If you instead damaged the lateral hypothalamus, the animal stopped eating and would waste away and die, unless force-fed. This led to the development of the 'dual-centre hypothesis'. Just like the sleep and wake centres in this area (see Chapter 5), this argued that the ventromedial hypothalamus was a 'feeding centre' and the lateral hypothalamus a 'satiety centre'. Turn on the feeding centre, or damage the satiety centre, and the animal would eat voraciously, unable to feel full. Do the opposite, and it would starve, without feeling hungry.

We now know that, as with much of neuroscience, this early theory was an oversimplification. While the hypothalamus is important for regulating feeding behaviour, there are other parts of the brain and nervous system involved too. But it seems the dual-centre hypothesis wasn't too far wrong. Rather than a simple hunger and satiety area in the brain, researchers now talk about hunger and satiety *networks*. Each of these goes through the hypothalamus but isn't limited to just one part of it. Stimulating one tiny region of the hypothalamus may make animals eat more, but that could be because it is having knock-on effects via that region's connections to other parts of the brain.

This system is also where the chemicals we discussed earlier have their effect. When insulin travels to the brain, for example, it inhibits our 'eat' circuit, while indirectly activating our 'don't eat' circuit, so we feel full. On the other hand, ghrelin, or an empty stomach, activates neurons in our 'eat' circuit, while inhibiting those 'don't eat' neurons. The balance between the two networks is delicate and can be tipped one way or another by a variety of inputs.

But finding the region of the brain that coordinates these signals still doesn't tell us *how* it regulates our appetite.

Clues to this came from those experimental rats with the damage to their hypothalamus. Surprisingly, they wouldn't continue to gain weight for ever. Instead, at some point they would slow down their eating and settle at a new, higher weight. If they were force-fed, they could gain even more weight, but once allowed to eat normally again, the extra weight would drop off. So the rats weren't simply always hungry. They were still controlling their weight, just at a much higher level than before.

This led to the development of a theory known as 'set-point theory', which argued that an animal's physiology works to maintain its weight at a set point. In 1953, Gordon Kennedy proposed the 'lipostat theory', arguing that fat deposits produce some sort of feedback signal to the hypothalamus, altering the amount of food an animal consumed. If it gained weight, it would reduce intake until it normalised again. If it lost weight, it would eat more to gain it back. He suggested the signal might be some kind of metabolic product, which would circulate in the blood.

Also in the 1950s, Romaine Hervey carried out a series of fairly grim experiments at the University of Cambridge which proved that this signal existed. This involved surgically joining two rats together when they were young, so they shared a blood supply. He found that when he lesioned the hypothalamus of one rat, it became obese, as expected. What happened to its partner though was more surprising. Rather than gaining weight, it began to lose it, eating very little even when it was offered food. He argued that this suggested a feedback mechanism, via the blood, which controlled eating. An animal would become insensitive to this signal if its hypothalamus was destroyed. The thin member of the pair was responding to this signal and trying (but failing) to reduce the weight of the pair as a whole, while the partner with the lesion munched on oblivious. This supported Kennedy's theory, but didn't get us any closer to understanding what the signal was. For that, scientists needed the help of two very special kinds of mice...

The fullness factor

In 1949, genetic research was in its infancy. But it was about to revolutionise our understanding of obesity and eating. Every year, the Jackson Laboratory in Maine bred millions of mice in order to better understand their genetics, and to supply other researchers with specific types of mice for their experiments. That summer, one of the animal caretakers found something surprising. One of the young mice was severely overweight: three or four times the size of its brothers and sisters. The scientists investigated and soon discovered this animal, and others they bred like it, had a mutation to a gene on chromosome 6. The gene was named 'ob' (for obese) and the mice, which had two copies of these mutations, became known as ob/ob.

In 1966, a different strain of obese mice, with a different genetic mutation, was discovered. These mice ate insatiably, were physically inactive, and suffered insulin resistance and diabetes, so were called db/db mice. Following in Hervey's footsteps, Douglas Coleman, working at the Jackson Laboratory, carried out a series of experiments on surgically conjoined mice, to try to find out what each of these genetic mutations was actually affecting.

When a wild type mouse (i.e. a mouse with no known genetic mutations) was connected to a db/db animal, it would stop eating and lose weight, just like the partners of Hervey's lesioned animals. And as before, the db/db mice munched on unaffected. This suggested that the db/db mice were *producing* a 'stop eating' signal, but they were unable to respond to it. When paired up, however, the wild type mice could detect this signal, and would stop eating in response. In contrast when ob/ob mice were connected to wild type or db/db mice, they ate less and lost weight, while their partners seemed unaffected. This suggested the ob/ob mice could respond to the circulating signal but didn't produce it themselves.

While these results were interesting, there was still no understanding of what exactly the 'satiety factor' that circulates in the blood could be, so it was met by some resistance by the scientific community. This resistance persisted right up until the point when the factor was discovered.

In the late 1980s, Jeffrey Friedman set up his own lab at Rockefeller University, where he had just completed his PhD. Since deciding to stop working as a doctor in favour of research, he had developed an interest in how chemicals in the brain could produce behaviour. It had recently been discovered that CCK, a chemical produced in the gut, also acted on the brain. Some researchers suggested this might be the cause of obesity in genetically obese mice, but others weren't convinced. Friedman was fascinated, so he began working with these ob/ob mice. As well as their overeating, these mice had a host of other problems, not seen in humans with obesity. They had low body temperature, problems with their immune systems, and were infertile. Friedman mapped the CCK gene, and compared it to the gene that was affecting the ob/ob mice. They found that the two were miles apart, on different parts of the mouse's DNA. So there was no way it was CCK that was the problem for these mice.

Friedman was determined to discover what the problem was for the ob/ob mice. Over the course of eight years, his lab used cutting-edge techniques to find exactly where on the mouse's DNA the ob gene was found. Once they had the region, they could examine all the genes in that area to see which was the most likely to be involved. Eventually, they discovered one gene in this region that was only 'turned on' in fat tissue. Cloning the gene proved that it coded for a hormone – they had discovered a chemical that could impact weight. They called it leptin, from 'leptos', the Greek word for thin. In an interview with the journal *Disease Models and Mechanisms* in 2012, Friedman recalled this moment of discovery:

> I think it was the singular most exciting moment of my professional life – it was completely overwhelming. The initial data showed not only that we had identified the ob gene, but also suggested that the gene product was under feedback control and probably encoded a hormone. This amplified the sense of excitement, because it was the first indication that Coleman's and our overall hypothesis was correct.

Now scientists had discovered the missing satiety signal, the idea of the hypothalamus as a regulator of body weight gained momentum. This discovery provided weight[*] to the set-point theory. The hypothalamus helps animals to maintain a steady weight by detecting leptin, which tells it how much energy we have stored in fat reserves. It combines this with information about the sugar levels currently available, circulating in our blood, and uses this to decide whether we should eat or not. If we overindulge at Christmas, and put on a few pounds, it tries to counteract that by releasing extra leptin and damping down hunger signals until the weight drops off again. And if we begin to lose weight, it encourages us to eat more. This is one of the reasons weight loss can be so difficult: our bodies fight against it, screaming out for us to eat more calories if our fat reserves begin to dip below our set point. With leptin levels lower, because of this reduction in fat reserves, it is hard to feel full, and so it's hard to stick to a diet. But in rare cases where humans, like the ob/ob mice, can't produce leptin at all, this problem can become extreme.

When Poppy was born,[†] there were no signs of what was to come. She seemed to be a healthy baby, and the birth was (relatively) easy. It wasn't until her parents took her home that they began to notice a difference between Poppy and her older brother, Evan. While their mother had been able to breastfeed Evan exclusively for the first six months, within a few weeks she realised this just wasn't going to work with Poppy. Her appetite seemed insatiable. Just half an hour after her last feed, she would be crying in hunger – so they began to bottle feed her.

---

[*] Pun intended!

[†] This story is based on a number of case studies of different patients with this condition, and the names are made up.

Her weight rapidly increased, and by six months old, she weighed 15kg (33lb) – roughly twice the weight of the average baby her age.

Her parents, who had both been slim all their lives, had no idea what was going on, but luckily Poppy was referred to an endocrine unit for assessment. Doctors there discovered she had a mutation to one of the genes involved in producing leptin, which meant that she wasn't producing the hormone at all. Doctors started her on replacement therapy, injecting the hormone, and within two weeks her weight started to drop. By the time she was three, she was no longer morbidly obese, and she was eating and behaving like a normal, healthy toddler.

Genetic mutations which stop individuals producing leptin are rare, but there are other conditions which can have a similar effect. Individuals with lipodystrophy, for example, have little or no fat tissue. This can be due to a genetic mutation they were born with, or it can be acquired later in life, but both forms lead to low circulating leptin levels. While they feel hungry and often overeat, individuals who can't make fat don't put on weight. But they do have a host of other symptoms. They often become diabetic, as they are unable to control their blood sugar and they can be resistant to insulin. They also have high fat levels in their blood, which can cause further problems. And they often have fertility issues. If these symptoms seem familiar, it is because they are very similar to those found in the obese mice, although these people were not obese. This led scientists to wonder whether the symptoms like infertility that the mice suffered weren't an effect of obesity itself, but down to a lack of leptin.

Throughout history, animals have evolved to survive short famines. We do this by storing energy in the form of fat, but animals also change their behaviour when they don't have anything to eat. To start with, they become hungry and more energetic, frantically searching for food. But if none is found, they enter a second stage. They reduce movement, and their body temperature drops, to conserve energy. The immune system functions less well, and the animal becomes infertile. Basically, the body does everything it can to conserve energy until the fast is over. During this time, leptin levels fall, and recently, scientists began to wonder if low leptin could be to blame for these symptoms of starvation. They tested it by giving animals in this fasted state leptin injections and, as predicted, found that the symptoms of starvation were reversed.

If leptin is our body's signal to our brain that we are starving, that explains a lot about both the mouse models and the cases of humans who don't produce leptin. Normally, our fat cells produce leptin, telling our bodies that we have energy reserves. If we lose weight, this circulating leptin level drops, and the starvation response is triggered, so we eat more to replenish them. But if you have no fat cells to make leptin, or these cells can't produce the hormone, or your brain can't respond to it, you would constantly be in starvation mode. Poppy's appetite was insatiable because she felt as if she was starving. And both the leptin-deficient and the leptin-resistant animals were the same; this explains both their overeating and their other symptoms. The symptoms of lipodystrophy also fit well with this starvation response explanation.

For individuals like Poppy, leptin therapy works. With her missing satiety factor replaced, her body realised she wasn't starving, and she was able to lose a large amount of weight. The US FDA (Food and Drug Administration) and the European Commission have now approved leptin as a therapy for lipodystrophy as well, after successful clinical trials, and the MHRA (Medicines and Health products Regulatory Agency) in the UK are considering doing the same. There have even been cases of women who are very slim (such as athletes or dancers) and who were previously unable to conceive, successfully becoming pregnant after being treated with leptin.

This, of course, sparks a question. Are all overweight people low in leptin? Could this treatment be the 'magic bullet' to help those who struggle to lose weight? Sadly, the answer is no. Studies have found that with a few rare exceptions, overweight people *can* produce leptin – often in large amounts. But what is common is for individuals to be resistant to the hormone, meaning it has much less of an effect on them. Friedman estimates around 10 per cent of people who are morbidly obese have a genetic mutation somewhere that means they produce lower levels of leptin, and for these people, treatment to raise levels has been shown to help with weight loss. But for the other 90 per cent, it would have no effect.

A weighty matter

Our understanding of how our bodies and brains control hunger and eating is increasing rapidly, but not as rapidly as our waistlines; 2.1 billion people worldwide are now estimated to be overweight or obese, and the numbers are still rising.

A quick caveat here: BMI (body mass index) is commonly used to calculate whether someone is a healthy weight. Unfortunately, it is a poor measure of health, because it uses only height and weight to make the assessment. As muscle weighs more than fat, very muscular people can come out as obese. Measuring body fat percentage would be better, but even this isn't perfect as where your fat is stored is important. Visceral fat, which is stored around your organs, is much worse for your health, so people who store fat on their thighs or buttocks are at lower risk for diseases like diabetes and heart disease than those who store it around their middles. In fact, someone may be a normal weight and appear slim, but still have unhealthy levels of visceral fat. And, of course, weight isn't the only marker of health; it is perfectly possible for a person of higher weight to be fitter and healthier than a leaner person who, for example, smokes, eats poorly and doesn't exercise. Unfortunately, BMI is still the measure used in most research papers, so when we discuss obesity throughout this chapter, it will have been measured this way.

There is also evidence that weight bias is harmful to the health of people who are higher weight. Shame and stigma can make people embarrassed to exercise, and can lead to increased eating and more weight gain. Some studies have even suggested that stigma can lead to poor health *independent* of actual BMI, perhaps because of increased levels of stress and the hormones associated with it. A study led by Angelina Sutin at Florida State University found people who experience weight discrimination were 60 per cent more likely to die than those who don't, independent of their actual BMI. There is also evidence of weight bias amongst healthcare providers and people with higher weight are less likely to visit a doctor, and often receive poorer care when they do, which might also explain some of the links between obesity and health problems.

What all this means is that the link between obesity and health problems may not be as direct as was once thought. But this area of research is still relatively new, and needs exploring a lot more. What is for sure is that throughout this chapter, when I talk about obesity as increasing the risk of disease, I am considering the population as a whole, not an individual. I am not implying the link is a direct cause. And I am certainly not saying that any health issues are the fault of the individual.

While acceptance may be important, for those who do want to lose weight, is there anything that can help? Weight loss is notoriously difficult, and your body has a trick to play on you. As we have seen, as someone starts to lose weight, their body responds to the loss of body fat as if they are starving, trying to maintain its set point by making them feel hungry. Being hungry has knock-on effects on your reward system (see Chapter 3), making food appear more tempting. These natural processes make sticking to a diet extremely difficult.

There is currently only one really effective treatment for obesity: bariatric (or weight loss) surgery. This can take a number of forms, but all aim to reduce the capacity of the stomach, so people feel full more quickly. This can be done by placing a silicone band around the stomach to squeeze it closed at the top (gastric band); putting a balloon inside to take up space; or cutting away part of the stomach (gastric bypass or gastric sleeve). After surgery, patients will feel full when eating very small amounts of food, restricting their intake and causing weight loss.

But this isn't the full story. People and rats who have had the surgery could just eat more often to overcome their smaller stomach – but they don't. This seems to be because rearranging the digestive tracts changes the release of hormones. After surgery, there is an increase in satiating chemicals like peptide YY, which activate the 'don't eat' network in the hypothalamus, and a decrease in the hunger hormone ghrelin. These changes work together to help people feel full and therefore eat less.

Studies have also found that gastric bypass increases the calories burned by rats, meaning they need 40 per cent more food than control animals to maintain the same body weight. We don't yet understand how this happens, but again, it might be linked to those neurons in the hypothalamus. The surgery seems to reset the set point of the animal, so its body then does everything it can to reach and maintain its new, lower weight, including burning more calories.

Patients who lose weight after bariatric surgery have been compared with similar individuals who lost weight through diet and exercise, and it was found the surgery patients had lower levels of the hunger hormone ghrelin. This could be why, in the long term, surgery is more effective than other weight-management plans. So, if these procedures are effective, at least partly because of changes to hormone levels involved in hunger and satiety, it might be possible to get this same effect without the surgery. While many procedures are now done via keyhole surgery, and are relatively safe, any operation carries risks, so drug treatment might be a safer alternative.

And it should certainly be cheaper. The science isn't quite there yet, but it is an exciting avenue of research.

For now, bearing in mind what we are starting to discover about stigma, there is a new approach being trialled. Known as 'health at every size', this puts the focus on increasing healthy behaviours, rather than losing weight. Evidence is emerging that, when compared to weight loss interventions, this approach is more successful at improving the health of individuals over the long term.

Food glorious food!

Hunger is important, but, as we can all appreciate, it isn't the only reason we eat, and perhaps it is these other reasons that are contributing to the world's increasing waistline. We've all had a second helping of dessert that we didn't really need, or nibbled through a packet of biscuits, just because they were there. So what is going on in the brain to cause this drive to eat even when, physiologically, we don't need to?

Gina Leinninger's group at Michigan State University is trying to work out the network of brain areas involved in overeating, and she has found a region of particular interest, the lateral hypothalamus.* This region connects to other areas of the hypothalamus, allowing it to receive signals from the body about whether you need food, but it also connects to dopamine neurons in the reward system. As we saw in Chapter 3, this system isn't so much for reward in the sense of pleasure, but is

* You may remember this region – it was the 'satiety centre' in the dual-centre hypothesis of eating we discussed earlier.

responsible for driving us to seek out certain things, or behave in a certain way, from taking drugs or having sex to eating. Leinninger explains:

> We know that dopamine signalling regulates the wanting of yummy, rewarding things... but it's also linked in with the brain systems that regulate how much we like them. Now how that gets coordinated with when we need to eat is usually through the lateral hypothalamus [LH]. The LH gets information from other parts of the brain that detect energy status and say 'hey, we don't have enough'. And the LH neurons then 'tickle' the dopamine neurons, to coordinate them to increase the motivation to actually go and get food.

For example, when blood sugar or leptin levels are high, the LH knows the body has plenty of energy reserves, so doesn't activate the 'wanting' system. As levels drop, however, it increases our desire to seek out food. This system is why when you are hungry, *really* hungry, food suddenly becomes all you can think about,* and when you do find some, it tastes amazing.

I experience this myself in hotels with a breakfast buffet. I'm a breakfast person – I usually wake up hungry and can never understand how others manage to make it to lunch on just a cup of coffee. I simply can't function until I've had something to eat. Usually, I'm pretty good at eating healthily. I have no problem passing up the cake after lunch for a bowl of fruit or eating a carrot mid-afternoon rather than cookies. But this all changes when

---

* I'm reminded of cartoons where the hungry character starts seeing their friends as a leg of ham...

I'm faced with a breakfast buffet. In that state of hunger, I find myself irresistibly drawn to the pastries – crisp, flaky croissants, tantalising Danishes filled with glazed apple and soft, fluffy muffins seem to call to me. That mixture of sugar and fat is like a siren song to my hungry brain.* To overcome this, I have developed a habit. Rather than picking and choosing my breakfast immediately, I don't allow myself an option. When I first approach the buffet I head straight to the fruit section, and load up my plate with fresh fruit and natural yoghurt. Only once I have eaten that, and my hunger is at least partially satisfied, do I approach the rest of the food. No longer 'starving', it is much easier to make healthy choices.

So the lateral hypothalamus is a regulation area, linking our needs with our desires. And as ever, this relies on chemical messengers. Mice without leptin signalling in this region overeat and become obese, and replacing leptin just in this area brings them back to normal. It also restores low dopamine levels. But leptin and dopamine aren't the only players in this region. There are also neurons that use a neurotransmitter called neurotensin. This chemical reduces the urge to eat tasty food, increases movement and suppresses weight. Leinninger believes that individual differences in how well it does this could explain why some of us are able to eat just one handful of crisps while others demolish the whole 'family-sized' bag:

> Disrupting the types of neurons that are meant to coordinate the dopamine neurons could lead to these differences. What makes this incredibly complicated to

* And, as we will see later, I'm not the only one who reacts like this.

> study is that there are so many different kinds of molecular, genetic factors in the lateral hypothalamic area and in the dopamine neurons, and any slight change in those could potentially change the way that this interaction functions and underlie that variability.

It may be that in obesity, this area somehow becomes disconnected from the internal state, so your motivation to eat is driven more by external cues like the sights, smells and sounds associated with food. With so many of these around, it's no wonder many of us are driven to overeat.

> If dopamine neurons or the LH became uncoordinated from that need for food, you might eat when you don't really need to. How this dopamine system gets disconnected, we don't know... But we think that dysfunction of the LH could be a driver.

Another brain area involved in this process is the prefrontal region. As we have seen, this area, behind your forehead, is involved in 'executive functions', like high-level decision-making and inhibitory control. Basically, this is the little voice that says you have had enough cake, even when you really fancy another piece. We don't know yet exactly how it connects into the rest of the feeding system, but we do know dopamine neurons play an important role, so it seems likely there could be links with the lateral hypothalamus here too.

Have you eaten?

As well as the obvious factors that influence when we eat, like hunger and being tempted by seeing something tasty, there is something perhaps more surprising that affects our appetite: memory. It turns out that memory plays a huge role in food choice. If you had pizza for dinner last night, your memory of that large meal may encourage you to choose a salad for lunch the next day. But it's even more fundamental than that. People with a type of severe amnesia, who can't form new memories, will eat a second lunch just 10 minutes after finishing their first one. They also don't report feeling any less hungry after a meal than they were before.

I don't know about you, but I find this really surprising. It certainly feels to me as if my sensations of hunger and fullness are concrete and visceral, rooted in my stomach, not my brain. I would previously have said that I eat less at dinner after a big lunch because I don't need the food. The fact that this is affected by memory throws doubt on what we think we know about ourselves and our bodies. If we can be wrong about the cause of our sensations of hunger and fullness, what other bodily sensations should we start to doubt? This is where science comes in. If we can't always trust our perceptions, we need to use scientific methods to dig down into what is causing them.

When we look at the connections within the brain in more detail, it begins to make sense. These severely amnesiac patients have damage to their hippocampus, a part of the brain that is vital for memories of things that have happened in your life, such as your 18th birthday, the last argument you had with your partner, or what you have just eaten (as we saw in Chapter 2). But this isn't the hippocampus's only role. This area of the brain also contains cells that receive signals about how hungry you are, via insulin and ghrelin receptors for example. This means it may also be involved in a process known as interoception.

Interoception is the ability to detect sensations from your body, such as heat, pain or hunger. For the amnesiac patients to feel just as hungry after a meal as before, they must not have been able to detect the signals coming to their brains from their bodies. And it turns out that even in those without brain damage, some people are better at interoception than others. This is tested by asking people to detect their own heartbeat, without feeling their pulse. People who can do this reliably are also better at detecting fullness when their stomach is stretched, and at feeling their stomach contract when it is empty. They may then rely more on these internal cues when deciding whether, and how much to eat, while other people, with less sensitive interoception, are more driven by environmental factors, like how tasty a food is, or the time of day.

Reading about this felt very familiar, as I feel like my eating is very much driven by internal cues. I am very bad at ignoring hunger, and always marvelled at the way colleagues could work through lunch if they had a tight deadline. For me, that internal signal is impossible to ignore. But I also can't overeat in one go, as the sensation of fullness in my stomach quickly becomes uncomfortable, something I find very frustrating when others are having that second helping of Christmas dinner! This means I am a grazer, eating little and often to keep my stomach happy. Perhaps some of these issues are to do with an overactive interoception.

In Marise Parent's lab at Georgia State University, researchers are trying to find out how the hippocampus, and its connections to the hypothalamus, control eating. To do this they are using a technique called optogenetics, which allows them to genetically manipulate rats so certain neurons can be switched on and off by light. This enables them to switch off cells in the hippocampus before, during or after a rat has eaten. They found that rats whose hippocampi had been turned off after their meal ate their next meal sooner and consumed around twice as much food. Just like the amnesiacs, it seems these rats don't remember that they have already eaten, and aren't able to detect the state of fullness, so they eat more. Interestingly, this effect occurs whether the rats are offered boring chow or delicious sugar, so it doesn't seem to be affected by how enjoyable a food is. It even works for saccharin, a sweetener that contains no calories.

And you don't need something as dramatic as amnesia or lights in your head for this to affect you. Research suggests that if you eat while playing a computer game or watching TV, you pay less attention to your food, and this is enough to encourage you to eat more later. Perhaps this explains the vast quantities of chocolate my husband seems to be able to consume while watching TV in the evenings, even after he has finished a big meal!

Being mindful about what you are eating, however, reduces future meal size. In fact, the amount of food you remember eating at lunch is a better predictor of how much you will eat at dinner than the actual quantity of food you consumed. So perhaps individual differences in the hippocampus, or the connections between it and the hypothalamus, could be to blame for some cases of obesity. Lucy Cheke, lecturer at the University of Cambridge, ran a series of studies that found that people with higher BMIs did fare less well on a number of memory tests. But this, of course, is a correlation, so we can't say for sure that their worse memory *caused* their obesity.

In fact, it could be the other way around. Studies have shown that feeding a rat a diet high in sugar and fat can actually impair their hippocampal memory, and studies in humans have suggested similar effects may occur in us. Whether because of inflammation, insulin resistance or other reasons, obesity seems to cause damage to the hippocampus and the rest of the memory network. In fact, one study by Kaarin Anstey and colleagues at Australian National University, following over-60s for eight years, found that the hippocampus of those with higher BMIs shrank more during that time than their lower-weight counterparts.

So perhaps we end up with a vicious cycle. Obesity causes memory impairments, which in turn make it harder for the individual to regulate their food intake, so they gain more weight. Finding a way to help people break out of this cycle could be a turning point in identifying treatments.

Stressed is desserts backwards

But even taking into account physiological hunger, psychological desire for food and mindless munching, we *still* haven't got the whole story about the factors that can drive us to consume.

Stress, for example, is a common reason people eat, particularly unhealthy 'comfort' foods. When we eat something delicious, our brains release chemicals that make us feel good. If we eat a doughnut when we are stressed, we feel better, for a little while at least. The problem is that many of us are stressed a lot. Our brains are very good at learning to repeat behaviours that have helped us in the past, so if that doughnut helped last time, it will encourage you to eat another the next time you are stressed. Rather than an occasional pick-me-up, comfort eating becomes a habit – and habits are hard to break. This is particularly true if the initial learning opportunity happened during childhood, when our brains are at their most flexible and changeable.

But what does it mean in terms of the brain to say we have developed a habit? That first time, we ate a doughnut because we were driven by a desire for a reward. Dopamine signalling in the nucleus accumbens made that food our goal, and we ate it. We experienced its deliciousness, and it made us feel better, so that behaviour (eating a doughnut) was reinforced. The brain has learnt it is a good thing to do when we are stressed. Over time though, as we repeat the behaviour, a different part of our brain takes over. The dorsal striatum becomes involved, and we now reach for that doughnut on autopilot, without actively wanting or enjoying it. Eating when stressed has become a habit.

But in the long term this self-medicating with tasty treats just might cause knock-on problems beyond the expanding waistline. Some argue that overeaters experience withdrawal symptoms when avoiding their favourite foods. Certainly, anyone who has been on a diet will have experienced how grumpy it can make you! The theory goes that denying yourself something delicious makes you anxious, and activates your stress system. You have learnt that food helps you cope with this state, so, of course, you are driven to eat to relieve the negative emotions. In fact, because of this system you may even end up eating more compulsively than if you hadn't tried to avoid the food in the first place.

And while it may make you feel better in the short term, some studies in animals found that consuming delicious food regularly, boosting your happy chemicals every time, actually desensitises the reward system in the long term. People with obesity do seem to fit this pattern, showing a reduction in dopamine receptors in reward areas of the brain, and a smaller response in this brain region when eating tasty food. This could lead to overeating, as an individual needs more and more food to produce the same sensation of satisfaction. And it may make you more prone to negative emotions in the long term, causing a vicious cycle.

If this is all ringing a bell, you wouldn't be the only one to think there might be similarities here with the drugs of abuse mentioned in Chapter 3. The same brain regions even seem to be important, as Gina Leinninger explained to me:

> LH [lateral hypothalamus] is important for drinking, for moving, for drugs of abuse, for pain – you name it! It seems to be the middle ground coordinator of [all sorts of] behaviours... We know that the dopamine system in general regulates reward for food, drugs, etc.; we don't know if every dopamine neuron does that the same way.

Just can't get enough

We've all seen headlines screaming that sugar, doughnuts, chicken nuggets or cheesecake are 'as addictive as cocaine'. But what does this actually mean? As we saw in Chapter 3, to be defined as an addiction, something must be pursued to the detriment of other things in a person's life.* But addictiveness is harder to define. It could mean how likely you are to get addicted after trying it once, or after a few uses. It could mean how likely you are to get addicted after long-term use, or even how likely it is that a person picked at random on the street is addicted.

You can see that these would give very different results. If we agree for a moment that sugar is addictive,† then if I picked someone at random, it would be more likely they were a sugar addict than a cocaine addict.‡ That's because most of us have more exposure to sugar than to cocaine. But if I gave 10 people a dose of sugar, and 10 a dose of cocaine, cocaine would probably come out as the 'more addictive' option.

On the surface at least, overeaters do share a lot of behaviours with drug addicts. Both groups feel driven to consume, even when there are obvious negative social or health consequences of doing so. Both feel cravings, and symptoms of withdrawal when they try to abstain. And both commonly experience relapses after thinking they had 'kicked the habit'.

But the debate is still ongoing about whether we should really think of food as addictive. In a recent paper,

* I had a friend at university whose desire for a jumbo sausage and chips after a night out was so strong that he would refuse to walk the girl he had spent the night dancing with home if it meant forgoing his midnight snack… which was certainly a problem for his love life!

† Which is far from clear-cut, as we will see later.

‡ Unless I was at a bankers' Christmas party in the City of London, perhaps!

presenting both viewpoints, the researchers first debate what it would be in the food that could be counted as addictive. Paul Fletcher, Professor of Health Neuroscience at the University of Cambridge, argues that you can't have an addiction without an addictive substance, leading Paul Kenny, Professor of Neuroscience at the Icahn School of Medicine at Mount Sinai, New York, to respond that it is not a single substance that is the problem but a combination of sugar and fat that would never be found together in nature, and 'pack[s] a supraphysiological punch to brain motivation circuits'. Fletcher, however, along with many other scientists, is as yet unconvinced by the evidence for sugar addiction, or sugar-plus-fat addiction.

There is certainly evidence, though, that our brains respond in a different way to foods containing a mix of fat and sugar than to either nutrient alone. When given access to just fat, or just sugar, rodents won't tend to put on much weight, even when allowed to eat as much as they like. Mix the two together, however, and they soon start to pile on the pounds. And humans too favour the mix of nutrients. In one study, carried out in Dana Small's lab at Yale University, Alexandra DiFeliceantonio and colleagues presented people with snack options, and asked how much they would be willing to pay for them. Despite previously rating the snacks as equally enjoyable, those that mixed fat and sugar (like a biscuit) were worth more money to the participants than those containing either mainly carbohydrates, or mainly fat.

Scanning their brains, the team found that activity in the striatum and other parts of the reward and motivation circuit was higher in response to these fatty, sugary snacks. But why would this be the case? One theory uses an evolutionary explanation. In our early history, foods high in fat or sugar were rare, and foods high in both were almost non-existent. And in a world where you don't know when your next meal might be, gorging on these treats when you do find them makes sense. But now, with these foods abundant, this same mechanism may drive us to overeat.

However, to really say that overeaters are food addicts, many researchers believe we should go a step further, and show that the same brain processes drive overeating as drive drug use. Many of the studies reported this way in the media try to do this, looking at brain activity when eating sugar, for example, and comparing this to taking a drug of abuse. And there are some striking similarities. As we have seen, eating sugary foods, particularly those that also contain fats, activates neurons in the reward system causing dopamine release, just as drugs do. This is what drives wanting, encouraging us to seek out more of that substance.

Tasty foods also seem to cause our brains to release natural pleasure chemicals, such as opioids and cannabinoids. These act on the same receptors as many drugs of abuse. But the quantities are important here; however tasty your chocolate cake is, it isn't going to produce the same quantity of opioids that a heroin user would experience. All this means that the debate is still open as to whether we should diagnose food addictions in the same way as drug addictions.

But there could be another way to look at the loss of control around food that some people experience, not as a substance addiction, but as a behavioural one. The inclusion of behavioural addictions in the DSM (the

Diagnostic and Statistical Manual of Mental Disorders, which lists all current possible diagnoses and is used by clinicians and researchers) has changed a lot over the years, but in the latest version, DSM-5, there is just one included: gambling. Interestingly, however, it has been given a new category: 'Non-Substance-Related Disorder'. This suggests there is room for more potential diagnoses in this category in the next version – perhaps Internet, shopping or sex addiction, all of which have been researched but haven't yet accumulated the evidence needed for inclusion. Maybe rather than thinking of addiction to a substance (food), we should look at this issue as a behavioural issue: eating addiction.

Tasty food is highly rewarding, and we saw in Chapter 3 that over time, dopamine release can be transferred to anything that predicts a reward. In the real world, this means the logo of your favourite restaurant, or the smell of freshly baked cookies. Experiencing these cues could drive you to seek the food as a habit, overriding any notion of whether you 'need' the food (how hungry you are) or whether you 'like' it. The wanting system takes control, in the same way drug cravings can be triggered by a place the user usually gets their fix, or gambling cravings can be triggered by walking past a bookie's. So does the same happen with food? There is evidence that people with obesity and binge-eating disorder *are* more sensitive to sights or smells in the environment that are linked to food. And how sensitive you are to these cues predicts weight gain over the long term.

While there are suggestions that there might be changes in the reward system in individuals with obesity, similar to those seen in sufferers of drug addition, it is early days for this research. As well as the reward system, there is also evidence that the prefrontal cortex is affected by drugs. This area acts as a brake on the reward system, helping to override it in cases where we shouldn't pursue the thing we want. If this brake is no longer effective, it can be harder to resist temptation. Some evidence hints at a similar effect in people who eat a lot of highly processed food, containing that heady blend of fat and sugar we like so much. But the jury is still out, and more studies in humans are needed to say for sure that the same mechanisms are behind overeating as addictions. There is also the question of whether it is a useful comparison. While it is possible to completely abstain from drugs, gambling or even sex, we need food to live. So treating it in the same way may not be helpful. But it certainly seems as if eating as a habit, rather than a choice, has a role to play in putting on weight. Interestingly, it also seems to have an impact on those at the other end of the scale: people with eating disorders.

What drives eating disorders?

At my all-girls secondary school, eating disorders weren't uncommon. Every year, it seemed, a girl would go missing from school. As days became weeks, the rumours would abound. Were they off sick with the flu, or was another student battling anorexia? When I discovered a close friend had been fighting the disease for months, I felt incredibly guilty. How could I not have noticed? But secrecy is one of the hallmarks of anorexia. I would see her eat at lunch, but not realise that with her active life, it was nowhere near enough. And, back then, I didn't realise that over-exercising was part of the disease. I certainly wasn't volunteering to go with her to jog around the field during break, but then I wasn't sporty, and she was. And her baggy clothes hid just how much of a toll the disease was taking on her body.

If overeating and its associated obesity is one end of the spectrum, anorexia is at the other. Despite being dangerously thin, most sufferers fail to see themselves as underweight, thinking they still need to lose more, and fearing weight gain. Untreated, they continue to lose weight, sometimes to the point of death. In fact, anorexia is the deadliest of all mental illnesses, and it is also one of the hardest to treat: 30 to 50 per cent of patients will recover with treatment, and tragically around 5 per cent die. The rest will struggle with the illness for their entire lives. My friend seems to have been one of the luckier ones; after some difficult times, she seems to be doing well, although we have limited contact now.

Our understanding of anorexia has developed considerably over the last 10 years or so. It used to be thought that the disease, which usually manifests during the teenage years, was an attempt to gain personal control in one aspect of a life that otherwise was controlled by others. It was blamed on the parents (or, more precisely, as is so often the case historically, on the mother). And it was blamed on the individual's personality and temperament – they were perfectionists, rule followers and highly anxious. Place this mix in a controlling environment and boom, anorexia. But more recent studies have shown this old theory isn't right. Yes, there are similarities in the personalities of people with anorexia, but this might just be a *result* of starvation, rather than the cause of it.

Interestingly, people with anorexia don't tend to restrict all calories equally. In a set of experiments, researchers including Karin Foerde and colleagues from the New York State Psychiatric Institute gave participants access to a buffet of foods and recorded what they ate. While healthy controls ate a wide selection of items, patients with anorexia tended to avoid those high in fat. Unsurprisingly, they also ate fewer calories overall.

The team followed up patients after treatment, when their weight was back at a healthy level. They had been successfully eating meals given to them by the clinic, which contained more calories and more fat, and were ready to be discharged. Worryingly, though, the researchers found the same pattern. Although the patients ate more calories than they had initially, they still, on average, ate less than the controls and avoided higher fat items. Their preference was still there, it had just been masked by not having free choice over their meals. This might be why so many people with anorexia, including my childhood friend, relapse, spending years in and out of hospitals and clinics. As soon as they are discharged, and begin making their own choices again, many return to their original pattern of eating. In fact, there seems to be a correlation between this avoidance of fat and likelihood of relapse. So tackling this might be vital in helping someone recover for good.

Another important factor is the variety of foods a patient eats. Studies including one by Janet Schebendach at Columbia University, and colleagues, have found that patients who eat the same thing every day during treatment, sticking rigidly to a set of rules, are more likely to struggle when out in the real world. This fits with a view of anorexia as a habit, put forward by Timothy Walsh, Professor of Paediatric Psychopharmacology at Columbia University. We have seen in previous chapters, and when talking about overeating, how behaviours can move from being goal directed ('I am going to choose that salad for dinner because I am trying to be healthy and had a big lunch, or because I fancy something fresh') to habitual ('I always have salad for dinner'). He believes this is what underlies anorexia. So perhaps an individual decides they want to lose a few pounds. They start a diet, and lose the weight. This feels good, so they repeat the diet. Over time, they find themselves trapped in a restrictive cycle. The restriction itself, rather than its outcome, becomes rewarding. Again, much like overeating, undereating becomes a habit, or even, perhaps, an addiction.

If you put patients in a brain scanner while asking them to make decisions about food, you can see this in action. In both patients and controls, regions of the prefrontal cortex are active, computing our individual valuation of each food.* But in people with anorexia, the dorsal striatum, responsible for habits, is significantly more active. The key to recovery, then, is breaking this habit. But this is never an easy process.

There are other factors involved too, like a person's response to reward. For most of us, delicious food is highly rewarding – it feels good to eat that ice cream, and we use this as part of our valuation of that food. But people with anorexia don't find it rewarding, as eating

---

* See Chapter 7 for more on how we calculate value and use it to make decisions.

fills them with anxiety instead of pleasure. We don't know exactly why this is, but there are theories. A number of studies have found people with anorexia are less sensitive to immediate rewards (like cake) and more sensitive to long-term rewards (like losing weight). They are also overly responsive to punishment, paying more attention to negative outcomes (like the possibility of becoming overweight) than positive ones.

Others have suggested that yet again, an imbalance in brain chemicals might be a root cause. Serotonin is involved in mood, and because of the link between low serotonin levels and depression (see Chapter 4) many people think boosting serotonin would make you happier. But the brain is a delicate machine, and while too little serotonin can be a bad thing, too much can be as well. High levels of this chemical have been linked with anxiety. Walter Kaye, a researcher who focuses on eating disorders, argues that people with anorexia have high levels of serotonin, a theory that is supported by the finding that they are more likely to carry a gene that increases levels in the brain. Serotonin is created in the body from tryptophan, an amino acid we get from our food. So, the theory goes, starving yourself is a good way to reduce serotonin. People with anorexia find that going on a diet, either voluntarily or because of another reason like illness, makes them feel less anxious, as their serotonin levels drop. They learn that they feel better when restricting calories, so they continue to do so.

But, as we have seen before, when you change the levels of a chemical in the brain, the brain fights back. To make the most of the little remaining serotonin, the brain of a starving person creates more receptors for the chemical. But that means that when they try to eat, or are forced to, the now hypersensitive brain is flooded with serotonin, overloading it and causing severe anxiety, even worse than before they restricted their food intake. This leads to a vicious cycle, with the individual 'self-medicating' through starvation, causing their brains to become more sensitive to serotonin and making it harder to restart a normal diet. The starvation itself has caused the disease to progress.

In testing this theory, researchers have found that people currently suffering from the disease do have lower serotonin levels in their cerebrospinal fluid and that these levels rise as they recover. Interestingly, even after long-term recovery, serotonin levels remain higher than in controls who have never had the disease, which may explain why for so many patients anorexia is a condition that must be managed for their entire lives.

The hope is that by understanding anorexia better we may be able to find a treatment, but sadly, so far, that has been elusive. Treating it like an addiction has limited success, and drugs that work for OCD (another idea was that it is essentially a food-related obsessive compulsive disorder) haven't been successful. They have, however, helped in another eating disorder: bulimia. People with bulimia go through bingeing phases, where they consume large amounts of food, and purging phases, making themselves vomit, or using laxatives or excessive exercise to counteract the eating. While this brings problems of its own, sufferers don't end up starved in the way people with anorexia do, so it may be that it is this starved state that prevents the drugs from working in anorexia.

Just like people with anorexia, individuals with bulimia may have disordered serotonin circuits. When we don't eat, for example overnight, our serotonin levels drop, but in people with bulimia, levels drop more dramatically than in healthy controls. Binge eating may be a way to try to restore these levels. Individuals with binge-eating disorder, who overeat like people with bulimia but don't purge, are thought to have chronically low serotonin levels, and again are trying to self-medicate with food. The link applies to those without a disorder as well; differences in a gene important for serotonin levels in the brain have been found to be associated with binge eating in the general population.

As well as serotonin levels, the reward system seems to be important in these disorders. While people with anorexia may produce too much dopamine, making them anxious and meaning they can go without normally pleasurable experiences like eating, bulimia is associated with low levels of dopamine and some of its receptors. When patients binge, dopamine increases in certain areas, relieving this problem. Those driven to binge seem to be extra responsive to rewards, and driven to seek them out. But this doesn't just go for food. People who binge eat are more likely to use drugs and alcohol as well. In bulimia, purging too is rewarding and patients with higher reward sensitivity are more likely to purge.

As we know, the reward system and the hypothalamus are tightly linked when it comes to eating. And Gina Leinninger's work on neurotensin (that chemical in the hypothalamus that suppresses eating) leads to an interesting possibility. Perhaps it is this that is dysregulated in anorexia. Interestingly, and perhaps counter-intuitively, neurotensin increases dopamine release, something that we would normally associate with an increased 'wanting', causing the animal (or person) to eat more. I asked her how more dopamine could reduce desire to eat.

> This is where things have gotten tricky… there may be different kinds of dopamine neuron… There has been evidence that activating some dopamine circuitry might actually reduce motivation to get rewards. Our theory is that these neurotensin neurons might engage a specific set of dopamine neurons and that their activation suppresses food intake. It seems to increase motivation to move around; it might be biasing the type of reward behaviour you go after. You want to go for a run instead of eating the biscuit.

These same neurons are the ones that decrease your desire to eat when you are ill, and put your hunger on hold when you are dehydrated, so you can focus on finding water. If these were disrupted in some way in anorexia, they could suppress the desire for food. Leinninger explained her theory: 'Disruption of the neurotensin system leads to this excessive dopamine signalling via these circuits that over-restricts the feeding. It short-circuits the brain – instead of being able to detect that we need food, it's overcompensating the food restriction.' She is currently applying for grants to study this, hopeful that if it does turn out to be part of the problem, it could lead to drugs or other interventions to help people with disordered eating at both ends of the spectrum.

This finding, that dopamine in a single area of the brain can have two completely different effects, highlights one of the issues in neuroscience research. The brain is incredibly complex, particularly when it comes to the balance of neurotransmitters needed for optimum health. Currently, the drugs we use to affect brain chemistry are blunt instruments, changing the amounts in all areas of the brain. This might have some benefits, but overloading our neural circuitry like this can also have unexpected downsides. Leinninger told me why she thinks so many of the drugs developed so far have failed:

> They may not have been ideally targeted. A lot of them are acting in many areas of the body… in the gut and in the brain… and have led to different responses… That lack of specificity has been problematic. By understanding how specific neural circuits work in this process there is hope that we could design treatments really narrowed down at the specific disruption in an individual… maybe even a personalised medicine approach.

Personalised medicine, prescribing drugs tailored to an individual, often based on their genetics, is just beginning to be used for some conditions like cancer. It is still a way off, but maybe one day, an individual's brain circuitry will be able to guide treatments. Leinninger is hopeful: 'Understanding the basic science of these processes will be able to help us develop treatments in the future.'

Let's all keep our fingers crossed she is right.

CHAPTER SEVEN

Logic, Emotion or Chemicals?

In Chapter 6, we discovered just how complex the brain architecture is that controls a seemingly simple decision: whether or not to eat. But every day, each of us has to make hundreds of decisions. Some of these are similarly simple, like what to wear or which route to take to work. But others are complicated and important, such as whether to take a new job or buy a house. We like to think that we make these decisions rationally, by weighing up the good and bad points of each option and coming to a conclusion. But that simply isn't the case. We can be influenced without even knowing it, and a lot of the time we base our decisions on emotion or assumptions, rather than cold hard evidence.

How we make decisions is something I have been interested in for a long time. I am not the best at choosing when it comes to mundane things. I find food shopping exhausting and get overwhelmed if a menu has more than a dozen choices. And I'm not alone here. Many people experience this 'decision fatigue', particularly when decisions are difficult because, for example, money is tight. Often, to stop the problem from taking over, I will just pick any option that is 'good enough', or ask someone else to pick for me.

My husband takes a different approach, trying to collect as much information on each option as possible (often in a spreadsheet for bigger decisions*), and taking his time to weigh up each option. This can be really helpful when it is a big decision like whether we should move house, but rather frustrating when it's a smaller one, like which bed to buy to furnish our flat. And when he goes food shopping in a new supermarket he can be gone for hours as he agonises over exactly which brand of muesli to buy! This led me to wonder why different people have these different approaches to decision-making, and how we can understand the process in general. And, if the networks for simple, binary decisions are so complicated, how on earth do our neurons weigh up multiple options, each with their own positives and negatives?

For what it's worth

To make a decision we have to, consciously or unconsciously, ascribe value to each of the options. If we are choosing between the vanilla ice cream or the chocolate, then the higher-value item is the one we would enjoy eating more. For most people this is a simple choice and mine would be chocolate, every time (the darker the better!). Sadly, in life, choices are rarely that simple. Say, for example, the vanilla ice cream was marked down, so it was half the price of the chocolate. Now I have to take into account the hedonic value of each choice (how much I will enjoy it) *and* the monetary value, so the decision becomes a more difficult one. Will I enjoy the chocolate twice as much as the vanilla? Is the

* Like which energy provider to go with – apparently a decision which warrants hours of research and multiple columns on a spreadsheet.

value of that extra enjoyment more important to me than the money I would save buying the discounted ice cream? How long can I stand in the frozen-food aisle staring at the freezers before someone escorts me out of the supermarket?

For decades, understanding these kinds of decisions was the domain of economists, and they came up with some theories. Traditional economics based itself on a set of assumptions, and used these to make predictions. They assumed we are all 'homo economicus' – a rational and selfish individual with relatively stable preferences. They also assumed we are able, and willing, to weigh up the pros and cons of each option, and come to a considered decision.

Let's take a gambling example. If we are offered the chance to pay £1 to enter a lottery where the chances of winning are 1 in 100 and the prize is £50, we should be able to work out that it isn't a good idea to play. We can do this using the expected value of the gamble, which is the average if we played it an infinite number of times. To get this, we multiply the value by the probability, giving an expected value of $50 \times 0.01 =$ 50p. So is it worth paying £1 for a lottery with an expected outcome of 50p? No!

But if this was the way we weighed up every choice in our lives, no one would ever enter the Euromillions lottery (except, perhaps, after a number of roll-overs) as the expected outcome is lower than the ticket price. It also can't explain why we insure our houses, when the chance of a fire is so low.

Here is another gambling game: I have a fair coin, and I'm going to flip it. To start with, I'll put £2 in the pot, but every time I flip a head, the amount in the pot doubles. The first time I flip tails, the game is over, and you get whatever is in the pot. So if I flipped tails the first time, you would get £2. If it's the second flip, it would be £4, and so on. How much would you pay to play that game? This is called the St Petersburg Paradox and it was initially devised by mathematician Daniel Bernoulli in the eighteenth century. If you work out the expected value, it is infinite.* But people won't pay an infinite amount to play. In fact, they won't pay much at all, which is why it was called a paradox – the theory just didn't match up with real life.

So why is this? Bernoulli suggested a few ideas. One hypothesis was that we simply discount very unlikely events, and so ignore the (tiny) possibility of winning a very large sum. Another one said it was because just looking at the expected outcome of a gamble is too simplistic. It assumes everyone places the same value on the same outcome, and that's just not true. So instead of expected value, he argued we should look at the expected *utility*. This is the individual's valuation of outcomes, rather than the outcomes themselves. He thought this could explain why people won't pay much to play his game. As you continue through it, if you win repeatedly, you quickly become incredibly rich. And now you are rich, the extra gains are worth less and less, so while the final value sums to ∞ (infinity), the final utility doesn't. Once you've got more money than you could spend in a lifetime, adding to that is, pretty much, pointless.

* $(\frac{1}{2} \times 2) + (\frac{1}{4} \times 4) + (\frac{1}{8} \times 8) \ldots = 1+1+1+1 \ldots = \infty$

Another idea is that we ignore the fact that the payoff is supposedly infinite, because we know that no one has infinite money. If we work out the expected value of playing this game against Bill Gates, where the maximum we could win is his 2020 wealth of $111.8 billion, the expected value is around $37 – no wonder people won't pay much to play.

So the expected utility theory is a bit better than the expected outcome theory. As well as your baseline wealth, it can also take into account some people being more risk-averse than others. But it can't explain why someone with £1 million who previously had £2 million will be less happy than someone who had £0.5 million and just got another £0.5 million. It assumes two people with the same amount of money should be equally happy, all other things being equal.

It also makes other assumptions of behaviour. For example, for these calculations to work, each person has to have defined preferences, and decide consistently, so if on Monday they chose chocolate over vanilla, and on Tuesday they chose vanilla over strawberry, you can know that given the option on Wednesday, they will choose chocolate over strawberry. But we all know our preferences can change from day to day, so this can't be true. It also assumes people are always able to decide, and never find themselves so blinded by choice they end up not picking any of the options, as often happens to me when I need a new bottle of shampoo. And it assumes that adding a third option won't influence your valuation of the two already on offer, whereas in reality it does.

Basically, traditional economic theories forget that humans are, well, human. We are messy creatures, influenced by our moods, the weather, and all sorts of other things. So, more recently, psychologists have stepped in, and created a field known as behavioural economics, to try to work out how humans *really* behave when faced with choices, and compare this to how the models predict we *should* behave. And even more recently, neuroscientists have got involved, to explore the underlying brain networks and chemicals that allow us to make these complex value-based decisions.

Wake me up before you no-go

As we saw in Chapter 3, dopamine is one of the ways the brain codes the value of something. When you do something good for the survival of the species, like eat tasty food, drink when thirsty or have sex, dopamine neurons in the ventral tegmental area (VTA) are activated. These neurons extend into the nucleus accumbens, increasing dopamine levels in this area and driving you to repeat that behaviour. So could it be this that allows us to make judgements about the value of various options, and decide between them?

Studies have shown that the average firing rates of neurons in the nucleus accumbens *do* correlate with the subjective value of an option, measured by how hard an animal will work to get it. This makes a lot of sense, as we saw previously that more dopamine in this area drives you to seek out a reward. But there is another layer of complexity here too, because this area can also take into account the likelihood of getting the reward. Say I was playing a strange ice-cream-based gambling game, where whether I win or lose is based on the roll of a die. If I play the vanilla game, I will get ice cream if I roll any number except a 6. If I choose the chocolate game, I must roll an even number to win. The salted caramel option only pays out if I roll a 1. Now my brain has something else to take into account, not just the value of each potential reward, but the likelihood of receiving it. And the dopamine neurons in the nucleus accumbens can do this too.

It seems neurons in this area code for expected value (or perhaps we should say expected utility). Imaging studies in humans have supported this finding from animal research. When people are playing gambling games, their nucleus accumbens is more active when they anticipate a large win. Interestingly, there do seem to be slightly different regions for value and expected value, but there is also a lot of overlap between them, and it is likely they work together to help us make decisions. So perhaps the economists were right, and we do make decisions based on the expected utility of the outcomes?

But, as is so often the case when it comes to the human brain, looking a bit deeper makes it clear that it's not as simple as 'more dopamine = choose that option', even within the nucleus accumbens. In fact, scientists now think that there are two pathways in this area, known as the 'go' and 'no-go' pathways. Similar to the networks that transition you from wake to sleep, or encourage you to eat or not eat, these different groups of neurons inhibit each other, and it is the balance between the two that determines if you move towards the reward ('go'), or stay away from it ('no-go').

Let's say it's early morning, and you are running late for work, so you dash out of the house before breakfast. At the station, you see you have five minutes to catch your train, but there is a queue at the coffee cart. Do you take the risk of missing your train in order to wait for a coffee and a muffin? Or do you skip the coffee and get on the train with a couple of minutes to spare? To decide you must weigh up the benefits of getting the coffee with the potential costs of missing your train and being late. Your 'go' pathway is motivated by the thought of the delicious coffee and muffin, the boost that the caffeine will give you and the feeling of hunger in your stomach. Your 'no-go' pathway is inhibiting this signal, acting as the 'angel on your shoulder' telling you it isn't worth the risk. And the balance between the two is driven by dopamine. When released, dopamine activates neurons in the 'go' pathway, and inhibits those in the 'no-go' pathway, tipping the scale. So a cue in the environment, like the smell of the coffee, might be enough to make you take that risk.

But our brains are clever, and they can learn. And, as we saw in Chapter 3, dopamine is important for learning too. When we predict how good something will be, dopamine is released in anticipation of the reward, driving us towards it. But if that reward is somehow worse than expected (maybe your coffee is cold, and your muffin is stale), this causes a prediction error. Our brains then use this information to learn, strengthening the connections between neurons, in this case in the 'no-go' pathway. Next time, you won't expect coffee from that cart to be as good, and that just might tip the scale towards 'no-go', and help you catch your train.

This dual system for decision-making can also explain why different people make different decisions, even given the exact same information. For example, there are the people who, when told to arrive at the airport two hours before a flight, add contingency time for queues at the car park, traffic on the road and any other possible eventuality, often ending up arriving well before check-in is even open (I must admit, I tend to fall into this camp!). Then there are those like my husband, who believe everything will be fine, and we should leave the house half an hour before the gate closes.*

So could it be differences in our go/no-go pathways that cause me to be more cautious, and him to be more willing to take risks? And which of us is actually making the better decision? Interestingly, it seems some people are driven in their decision-making more by the go pathway and others are guided more by the no-go pathway.

To study this, Michael Frank, then at the University of Colorado Boulder, developed an experiment. Participants saw pairs of symbols, and had to choose one, before being told whether they were correct. But it wasn't a one-to-one relationship; symbol A wasn't always correct and symbol B wasn't always incorrect. The relationships varied, but in the most reliable pair, A was correct 80 per cent of the time. Other pairs (e.g. C versus D) were more random. Over time, participants learnt which symbols were more likely to lead to a win (simply being right is enough for our brains to count it as a reward),

* OK, that might be an exaggeration, but you take my point!

and which were less likely. But there are two ways to do this: you can learn either to choose A, as it leads to reward more often, or to avoid B, which leads to punishment (being wrong) more often. To tease these apart, once they had learned these relationships, the pairings were changed. Now, for example, rather than always appearing as A versus B or C versus D; symbol A might be paired with symbol D. Would participants given these new pairs prefer to pick the 'good' option, always picking A whatever it is paired with? Or would they be better at avoiding the 'bad' option of B?

To look at the role of dopamine in these choices, the team investigated people with Parkinson's disease. The damage to dopamine neurons caused by this illness leads to problems with decision-making, as well as the characteristic difficulty with movement. But previous findings had been confusing. In some cases, dopamine-boosting drugs improved their decisions, while in others, it made them worse. Frank's theoretical work suggested to him that dopamine might be having different effects on different networks of neurons coming from the striatum, the area of the brain that contains the nucleus accumbens. As he predicted, his team found that people who weren't taking medication were very good at learning to avoid B, but found it harder to learn to pick A. Their 'no-go' system was working, but 'go' was impaired. Taking their meds, which increased their levels of dopamine, reversed this, tipping the balance towards the 'go' pathway.

They also tested healthy volunteers given drugs to raise or lower dopamine levels. As Frank writes in an article for the Dana Foundation:

> We found a striking effect of the different dopamine medications on this positive versus negative learning bias... While on placebo, participants performed equally well at choose-A and avoid-B test choices. But when their dopamine levels were increased, they were more successful at choosing the most positive symbol A and less successful at avoiding B. Conversely, lowered dopamine levels were associated with the opposite pattern: worse choose-A performance but more-reliable avoid-B choices. Thus the dopamine medications caused participants to learn more or less from positive versus negative outcomes of their decisions.

Could this explain the difference between myself and my husband? Could he be more of a positive learner, while I learn more from negative outcomes? And could this be linked to dopamine levels in these pathways? While we can't know for sure, it is possible. Studies on healthy volunteers have shown that there are genetic variants linked to dopamine in this area, which can predict some of these differences between individuals. As for which pathway leads to better decisions, it seems, as is so often the case, that balance is important. If either pathway is dominant over the other, it can lead to poorer learning and poorer decisions.* The brain must maintain a delicate balance of dopamine to allow the competing systems to work most effectively and appropriately. Perhaps there is a metaphor here for balance and compromise in marriage too!

---

* Although in the real world it is much harder to define what is a 'good' or a 'bad' decision than it is in the lab.

Going with your instincts

So we know that the striatum and its dopamine neurons play an important role in decision-making. But, if we have learnt anything throughout this book it's that in the brain, things are rarely that simple. The striatum might be there to weigh the possible benefits of two options, but we all know that decisions aren't based just on potential benefits. We have to take into account the potential costs and downsides of each option too. And our dopamine neurons, focused on reward, aren't very good at doing this.

In a set of experiments, Mark Walton at the University of Oxford and colleagues at the University of Washington trained rats to press levers to receive food rewards. Sometimes there were two levers in the box so the rats had to make a choice, while at other times only one was available, forcing their decision. This meant the researchers could find out a rat's preference, but also look at its brain responses when making the preferred and the non-preferred choice.

As well as varying the amount of food the rat received with each press, they could vary the number of presses it had to make to get that food. So, for example, they might train a rat that the left lever gives one food pellet for four presses, but the right one gives six food pellets for twenty presses. Does the rat work out that the harder option is the better cost/reward ratio? And when you find a combination that gives a clear favourite, what is the dopamine doing in its brain?

Interestingly, it seems that rats do have preferences, based on the ratio of work to reward. But it isn't coded by dopamine in their nucleus accumbens. In another task the researchers compared an easy option that gave a small pay-out with a harder option that gave a slightly larger one. The rats consistently preferred the easy choice, working out it was better value. But their dopamine cells still responded more to the bigger of the two rewards, even though they had to work much harder for an only slightly bigger gain. This tells us that while dopamine does code value, it is mainly associated with the worth of the reward, barely accounting for the effort needed to get it. But this isn't how we, or rats, make decisions. So there must be something else going on to take negatives into account. And this is where our emotions come in.

Our emotions are controlled by a complex network of brain areas (or possibly multiple networks) commonly known as the limbic system (see Chapter 4). While we all know how emotions feel, defining what they are is challenging, but at a most basic level some researchers have argued that they are states elicited by rewards and punishments. And, as such, it is easy to see why they should play into our decision-making. If a decision you made produced a reward, making you happy, you should do it again. If a choice caused punishment, or the removal of that reward, making you sad or angry, you should avoid that decision in the future. This type of emotional learning relies on the amygdala and is hugely important for making good decisions.

When we are presented with multiple options, there are different ways we can go about making a decision. We could pick by chance, which is quick and easy, but usually won't give us the best outcome. We could make a rational decision, weighing up the pros and cons of each option (like my husband) but that takes a long time, and a lot of effort. Or we could rely on what we call heuristics – quick, rough approximations that give us a good answer more often than not, and are easy to use. These, some argue, are what emotions are. Trying to choose a restaurant, and one of the options brings back a negative emotion after you came across a rude waiter there? Easy, rule it out. Get a gut feeling about a particular sofa when out shopping for one? Great, done!

It may not sound like the best way to make complex decisions, but experiments have shown that your emotions are a valuable guide, and if they are blunted somehow, it can lead to disastrous outcomes.

Pick a card, any card…

One famous experiment used to look at emotional decision-making is known as the Iowa gambling task (IGT). A group of researchers at the University of Iowa, including Antonio Damasio, devised a game where players are presented with four decks of cards. They must choose a card from any deck, and turn it over to reveal if they have won or lost money. Unbeknown to the participants, two of these decks of cards are 'good decks'. On average, players using only these decks would accumulate money over the course of the game. The other two are 'bad decks' which, on average, will cause the player to lose money. But the 'bad decks', while losing you money in the long run, also have some big wins in their stacks.

When healthy subjects play this game, they quickly learn to focus on the good decks and avoid the bad.

Over the years since this experiment was devised, it has been repeated on a huge range of patient groups and healthy volunteers, and different factors have been measured. In one of these experiments, Damasio and colleagues discovered that players with damage to their amygdala don't seem to be able to learn this rule.

To find out what was going on, the researchers measured the players' skin conductance to tell whether their emotions were heightened.* Normal subjects experience an increase in skin conductance when they get a reward or a punishment and over time develop an anticipatory increase, with their skin conductance increasing before they choose a card, especially when they consider choosing a bad deck. People with amygdala lesions have much lower responses to reward and punishment, and don't develop this anticipatory response.

These findings provided support for Damasio's somatic marker hypothesis, the idea that we use emotions to help us make good decisions quickly. In this gambling task, the authors argued, we learn to associate the bad decks with the negative feeling we get when punished. People with damage to their amygdala can't do this, because the amygdala is essential for emotional learning – it is what attaches the emotional response to the object. These people did show an emotional response to a loud noise, so it wasn't just that they couldn't feel emotions, but they couldn't learn to associate that sound with something

* When you experience a strong emotion, good or bad, your hands start to sweat, ever so slightly. You might not notice it, but it makes it easier for electricity to be conducted over your skin. By placing sensors on the hand, scientists can measure this change, and use it to detect emotions you might not even know you were feeling.

that predicted it, as healthy people rapidly do. And this is backed up by animal studies that have found destroying the amygdala prevents learning about things relating to fear. So our amygdala allows us to use emotions to make good decisions. But it isn't the only region involved. Interestingly, the first study using the Iowa gambling task wasn't looking at the amygdala, but an area of the frontal lobe known as the ventromedial prefrontal cortex, which sits just above the sockets of your eyes.

Many of the areas in the brain network vital for decision-making are found in our frontal lobe. As we have seen in previous chapters, this area is known for controlling our most 'high-level' functions, such as reasoning, planning and cognitive control. So it isn't surprising it is important for making decisions. Damasio and his colleagues were interested in the ventromedial prefrontal cortex because of the difficulties faced by people with damage here, particularly a patient known as EVR. By the time he was 35, EVR was married with two children and had worked his way up to be responsible for all the accounting operations of a home-building firm. He was active in the church and his four younger siblings said he was someone they had always looked up to. But then things began to change. He had problems with his vision, and he began to act differently. It turned out EVR had a brain tumour in both of his frontal lobes, so surgery was scheduled to remove it. After the operation, he seemed to recover well, and was discharged from hospital two weeks later.

When it came to follow-up appointments, however, it was clear that all was not well. Three months after his surgery EVR started work again, but soon began a partnership with a former co-worker, who had been fired by the company. Friends and family warned him that this man wasn't to be trusted, but EVR ignored their pleas and invested all his money in the venture, which soon failed, leaving EVR bankrupt. He managed to secure another job but was quickly fired because despite having the required skills he was disorganised and often late. His marriage of 17 years also broke down, leading to a divorce. EVR was forced to move in with his parents, and a month later he was married again, this time to a prostitute, despite his family's protestations.

Follow-up tests confirmed that the tumour had not returned and showed that his IQ remained intact. In fact, he scored highly in all the tests given to him, including for memory and personality. But something wasn't right. While he could hold an intelligent conversation about world events and financial matters, EVR could be completely overwhelmed by trying to make a simple decision about where to have dinner or which shirt to buy. He became stuck in never-ending loops, trying to weigh each aspect of each item rationally, even driving to restaurants to see how busy they were before, often, making no decision at all. Still his tests came back normal. As neurologist Paul Eslinger and Damasio explain in their 1985 paper about EVR: 'The conclusion was that his "adjustment problems" are not the result of organic problems or neurological dysfunction… instead they reflect emotional and psychological adjustment problems and therefore are amenable to psychotherapy.'

But therapy didn't help, so EVR ended up being treated by Eslinger and Damasio. Again, they put him through batteries of tests, all of which he performed well in, even those designed to test frontal-lobe functions. However, imaging studies of his brain showed damage to the ventromedial prefrontal cortex, on both sides of the brain.

One of the most striking results was that EVR could answer questions about social situations appropriately, but behaving that way in a real-life situation seemed beyond him. As the researchers put it:

> As he awoke, there was no evidence that an internal, automatic program was ready to propel him into the routine daily activities of self-care and feeding, let alone those of traveling to a job and discharging the assignments of a given day. It was as if he 'forgot to remember' short- and intermediate-term goals.

This was not impulsiveness, or reckless behaviour, as seen in people with damage to other areas of the frontal lobe. Nor did he struggle to initiate movements. It was only complex sequences of goal-driven actions that he seemed to have a problem with. The scientists realised that it was the areas of the frontal lobe that connect to the limbic system that were affected, so they thought this could be the source of EVR's problems. His emotional areas and rational prefrontal areas could no longer communicate, in either direction. This meant, the authors wrote, that he could no longer regulate his actions properly. On the one hand, his frontal lobe couldn't modulate the more basic drives, based on his environment, and in the other direction, those basic 'drives and tendencies' couldn't activate his frontal regions so vital to making the decisions needed to follow through on goals.

It was an attempt to understand the problems suffered by EVR and people like him that led to Damasio and colleagues devising the Iowa gambling task. And, as they had anticipated, EVR showed a specific impairment in the task. Just like the patients with amygdala damage, people with damage to their ventromedial prefrontal cortex, like EVR, made risky decisions, continuing to pick from the bad decks. But there were some interesting differences between the two groups when it came to their galvanic skin responses. Unlike people with amygdala damage, those with damage to the ventromedial prefrontal cortex did respond emotionally when they won or lost money. But they didn't develop *anticipatory* responses, as the healthy participants did. Fascinatingly, many of these patients were able to tell researchers which decks were 'good' and which were 'bad', but this didn't stop them picking the bad ones over and over again. In healthy participants, the increase in skin conductance before picking the 'bad' deck started to appear before they could consciously state which deck was good and which was bad, but this 'hunch' phase seemed to be missing for people with damage to the ventromedial prefrontal cortex.

So why the similarities and differences between people with amygdala damage and people with damage in the frontal lobes? These findings suggest both regions are part of a decision-making network, but that they play different roles. People with damage to the ventromedial prefrontal cortex do experience emotions when winning or losing and, unlike people with amygdala damage, they can learn to associate a cue with an emotional event, like a loud noise. But they can't *use* these emotional experiences to guide decision-making. During the task, each deck will be associated with both wins and losses, and so with both positive and negative emotions. People with damage to the ventromedial prefrontal cortex can't integrate all this information and use it to make a decision. They don't recreate the emotional states associated with a deck before picking a card, as healthy players do, so can't use these 'somatic markers' to guide them. And as a result, they make bad decisions.

In real-life contexts, while people with damage to the amygdala have no sense of danger and will make choices that will directly harm themselves and others, people with damage to the ventromedial prefrontal cortex like EVR don't make that kind of bad decision. Instead, their decisions tend to cause problems down the line. Damasio and his team interpret this as supporting their hypothesis: people with damage to the ventromedial prefrontal cortex made poorer decisions because they couldn't use somatic markers to guide them. It seems not re-experiencing the emotions associated with past mistakes means they are unable to learn from them.

Chemicals in control?

This research is fascinating, and tells us a lot about the brain networks involved in decision-making. But this is a book about brain chemicals, so of course our next job is to find out whether certain neurotransmitters have been linked to this process of emotional decision-making. And it turns out Damasio did put forward some ideas, albeit tentatively. As we saw in Chapter 4, emotions are changes in the body and the brain, and many of these changes are chemical. For example, the bodily state experienced when losing a lot of money is likely to involve stress hormones like adrenaline and cortisol. Damasio thought that while the re-creation of these states before decisions might not produce the same bodily changes, it could trigger chemical changes in the brain, particularly highlighting dopamine and serotonin as mediators of this signal. At the simplest level, changes in the release of these chemicals in one area of the brain could change how easy it is for another area to be activated.

We know dopamine helps the brain to weigh the importance of each option, so this makes a lot of sense. Emotionally triggered changes in the levels of this chemical could boost the attention and importance we place on one response over another, helping us make the decision.

Serotonin's role has also been studied, although results are far from clear cut. One way to explore how serotonin affects decision-making is to give participants SSRI drugs, which boost levels of serotonin in the brain. Some studies have found this is beneficial, increasing the number of 'good' selections in the Iowa gambling task, especially in the second half of the experiment, when the subjects at least have a hunch about what is going on. But the jury is still out – other studies haven't found the same result.

Another way to change serotonin levels in the brain is by increasing or reducing the amount of the molecule it is made from, tryptophan. Giving humans extra tryptophan, for example, can reduce bias in decision-making. And depleting it, so reducing serotonin levels, does seem to impair decision-making in rats, adding support to the idea that serotonin is involved in the process.

We can also look at natural variations in levels caused by genetic differences. There is a gene, for example, called 5-HTT that controls the serotonin reuptake transporters. These clear away excess serotonin from the synapse, stopping it from continuing to activate the post-synaptic neuron. As we saw in Chapter 4, it is this protein that SSRIs target. Blocking it means serotonin hangs around in the synapse longer, having more of an effect. Completely knocking out this gene in rats boosts the serotonin available in the synapse and improves performance in a rat-version of the IGT. But in humans, the story is, as ever, more complicated. In the human population, we can't do this exact experiment, but we can compare people with different variants of this gene, which affect how rapidly the serotonin is hoovered up.

If you have the short version of the gene, your transporters are a bit more lax with their cleaning process and let the serotonin build up in the synapse, like a student allowing washing-up to accumulate in the sink. And, the theory goes, this should be beneficial for decision-making. Some studies have found this, with subjects with the short version of the gene doing better in the IGT, but others have reported no difference, or even a worse performance for these people. Studies in patient populations have been no clearer. This could be down to differences in the methods used – whether the studies look at the IGT as a whole, for example, or broken up into sections. They also haven't looked at *how* serotonin is having its effect, and whether it relates to the somatic marker hypothesis.

A group of researchers in Romania, including medical geneticist Vulturar Romana, report a study that tried to tease apart these effects, looking at whether there were differences in the emotional responses of people with the lazy transport proteins (the short genes) versus the hyper-efficient ones (the equivalent of that person who tidies away your mug before you have quite finished your cup of tea*). They did find a difference. The low-functioning group had higher emotional responses, and made better choices than the group with hyper-efficient proteins. And, they argued, this was a causal relationship with the increased serotonin available to the low-functioning group giving them more dramatic emotional responses and causing them to make better decisions. In fact, they found that this relationship could explain nearly half of the difference that the genes had on outcomes in this experiment. But again, it seems likely that the relationship isn't as simple as more serotonin leading to better decision-making overall. In the Iowa gambling task, using your emotions to make the decision is helpful, but there are times in our lives when emotions can get in the way of making the best decisions, like avoiding a restaurant you love after having an embarrassing experience there. And other studies have found the group with the short version of the gene are more risk averse, and more prone to biases. They are also more anxious. So it could be that their stronger emotional

* Yes, I'm looking at you, Mum!

responses, while helpful in some decisions, can hamper them in others.

Can't see the orchard for the apples

It seems our genetic make-up has an impact on our decision-making, but even when it comes to a single individual, we humans can't always be relied upon to make the best decision, or even to make the same decision twice. One way to think about our decision-making selves is as two people, battling it out for dominance. There is our emotional system, primarily driven by dopamine in the nucleus accumbens, crying out for reward. It's a bit like a toddler, and once it has set its sights on that piece of cake, it's going to throw a tantrum until it gets its way. Then there is our reason system, based in the prefrontal cortex – the harried parent, trying its best to keep the emotional toddler under control. But as with many parents, when our brain is busy, or stressed, or tired, it can't always face saying no to the screaming child. Sometimes it is easier to give in, and just give it what it wants.

Stress can affect the brain in many ways, and it's important to distinguish between chronic and acute stress. Someone who experiences high levels of stress for a long time, particularly from childhood, will have their brains shaped by this (see Chapter 2). Long-term stress can reduce the functioning of the prefrontal cortex and may also change an individual's response to reward.

But even short-term stress, of the kind we all experience regularly, can change the way you make decisions. To test how this works, researchers need a way to generate stress in the lab that is consistent and reliable. Rather than making participants read the news (which may currently generate anything from anger to despair!), they ask people to plunge their hand into a bucket of iced water and keep it there throughout the task. This may not sound too bad, but it is seriously unpleasant, and gets more so the longer it goes on. And we know that it causes a lot of the same physiological changes that we see when someone is stressed in the more common meaning of the word, by being overworked or having money worries.

In one task, Jennifer Lenow, then at New York University, and colleagues gave participants a game to play while they were holding their hand in cold water. They were shown an orchard on a computer screen, and asked to collect as many apples as possible. On each trial they could either continue to pick from the same tree, which would yield fewer apples as the trials went on, or move on to the next tree. The orchards came in two types: rich, where the trees were close together, and harsh, where it took longer to travel between them.

The researchers found that participants chose to move to the next tree sooner in rich environments, as expected. But when they compared the stressed group to the control group, whose arms were in warm water, they found a difference. The more stressed a participant was (measured by self-reports as well as cortisol responses), the longer they stuck it out at the same tree, before moving to the next one. And this was true in both rich and harsh environments. When compared to the optimal point to move (calculated by the researchers), stress was associated with overexploiting the current tree.

The researchers argued that stress biased participants' perceptions of the environment, making them think it was harsher than it was, so encouraging them to stay put. But there are other theories. Studies have found that stress can make people more likely to pay attention to positive rather than negative information. So in this case, they were focused on the fact that they were getting some apples (hooray!) rather than the fact that there were fewer than there had been (boo!). Stress also seems to bias people towards inaction. So when stressed we are more likely to stick with the status quo (the current tree) rather than take action and move to a different one. In fact, stress impairs learning specifically when an action is required, whether that action leads someone towards a reward or away from a punishment. Another finding is that stressed people think about the present more than the future, potentially making it harder to see the possible benefits of moving to a new tree.

One of the ways stress could affect decision-making is via the dopamine system, and as we saw in Chapter 4, studies have found changes in this system after both acute and chronic stress. But exactly what changes are seen varies depending on the type of stress. For example, one study on mice found that chronic mild stress reduced the activity of dopamine neurons in the ventral tegmental area, and stimulating these made the mice behave as if the stress had been reduced. However, when stress arose from a social task, the dopamine neurons fired more, and other researchers have found that an acute stress like an electric shock can also increase dopamine neuron activity in this area.

So how can this be? How can stress both increase and decrease the activity of these neurons? One answer may be that the resolution of these studies wasn't good enough. As we have learnt, there can be different populations of neurons, even within the same area, which can do very different things – the 'go' and 'no-go' pathways for instance. Maybe different types of stress affect different networks of neurons within the same brain area. Or perhaps it is a case of the brain fighting back, changing its response when stress is repeated or ongoing. More experiments are needed to find out for sure how stress affects dopamine, and how this impacts our decision-making.

While we may not quite know how yet, it does seem stress encourages animals to choose the easy option, rather than an option that involves working for a bigger reward. In humans too, stress often seems to push our decision-making towards the more instinctive responses, rather than the deliberate. This makes sense, as these heuristics and biases have evolved to guide us and help us to quickly and easily make the decisions that will help us survive. Whether stress has a positive or negative effect on our decision-making varies depending on the type of decision, and from person to person.

This is, at least partly, because we all respond differently to stress. Some people are calm and laid back normally, but ask them to speak in front of a crowd and they become a gibbering wreck. Others, like me, love the limelight, but begin to feel panicky just thinking about a tight deadline – something many others thrive on.

To measure someone's stress level, we often use the amount of cortisol in the bloodstream or the saliva. While cortisol isn't the only chemical involved in stress (see Chapter 4), it is a good way to measure how an individual is responding to whatever type of stress they are under, although you do have to account for the fact that cortisol levels vary with time of day.

When cortisol levels are high, it binds to receptors in the limbic system, and this could be how it changes decisions, by altering the connections between prefrontal and limbic regions. In animals, it also seems to boost dopamine signalling, making the animals more driven by rewards, and more likely to take risks to get them. A study where humans were given a tablet containing cortisol supports this finding, as after they had absorbed it, participants were more likely to choose a riskier gamble than controls. It must be noted, however, that this was a study in young men, a group more likely to take risks than any other. As with many psychology studies, this means we can't know for sure that the same would be found in older people or women, and other studies have suggested stress might affect these different groups in different ways.

One big problem with our increasingly stressful lives is that stress can make us more susceptible to biases. These can range from the 'self-serving bias', which means that 90 per cent of us would rate ourselves as 'above average' on anything from being a friend to driving, to the implicit racial or gender biases that we all show, whether or not we choose to act on them. But knowing this is half the battle. Once we are aware of our unconscious biases, and the states in which we are more likely to fall prey to them, we can begin to counter them. While there just isn't room in this chapter to go into all the biases we humans are susceptible to,* I have picked one to look at more closely.

Show me the money!

Sometimes, when we make a decision like which breakfast cereal to buy, or which movie to watch, we compare a range of options that are in front of us. But a lot of the time, the options aren't concrete, immediate things, and the payoff for one of them might be far into the future. Should you, for example, eat another slice of pizza? The short-term benefits of its deliciousness are obvious, but the long-term consequences of the extra salt and its impacts on your blood pressure could be dire. Should you stay in your boring but stable job, or quit to start your own business? Buy a new top, or save that money so you can go on holiday later in the year? Marry the person you are with, or wait around in case someone better comes along?

In real life, these are the kinds of decisions we constantly have to make, and to do this, we must put ourselves in the shoes of our future self and imagine how our decision will impact them. Unfortunately, we are pretty bad at this. In some ways we have a very high opinion of 'future us'. Given a choice of snacks for next week, for example, we will tend to go for the healthy option, because future us likes to look after themselves. When the time comes around though, the chocolate bar starts to look more and more tempting. We will also

* Wikipedia lists over 120 biases relating to decision-making alone.

volunteer more hours of time for a good cause when that time is in the future, and will sign up to watch a highbrow foreign movie next month, whereas tonight we just fancy a silly rom-com. Our future self is the person we would like to be, generous, cultured and healthy. But we also find it difficult to relate to them and imagine what it would be like to be them, forgetting that they probably like chocolate or rom-coms too.

This is partly due to something called temporal discounting, and it can lead to some major problems, such as the fact that over a third of UK adults don't have a pension. Even those who do aren't saving enough. Only 28 per cent of people asked by comparison website Finder believe they are on track to save enough for a comfortable retirement.

Temporal discounting is a term for the fact that people favour smaller immediate rewards over larger delayed ones. We can do a simple experiment to see how this works. Imagine I offered you £10 now or £15 tomorrow – you would probably wait for the larger amount. But what if I offered you £10 now or £15 in a year's time? The immediate option starts to look more appealing. Experiments like this in both humans and animals allow us to produce discounting curves, showing how the value of a reward drops off as it is pushed further into the future.

Temporal discounting may have evolved as a beneficial strategy. After all, who knows if you will be alive to get a bigger pay-out in a year, or if the person offering it will! But in the modern world, where most of us *do* make it to old age, it leads to some of the issues facing society today, where the majority of people don't manage to save enough to cover their needs as they get older, at least partly because the pleasure of spending the money now is just too tempting.*

As well as behavioural measures, if we look into an animal's brain, we can see the neural basis of this discounting. Once an animal has learnt that a cue predicts a reward, you can change the delay before the reward arrives, and see what happens. And it turns out that the more delayed the reward, the less active the dopamine neurons in the nucleus accumbens are. This area doesn't just take into account a reward's size and probability but also its timing, combining these into that signal of subjective value.

Perhaps surprisingly though, some areas of the prefrontal cortex don't differentiate between the timings of a reward, and others are more active when someone chooses the delayed option. This brings us back to that idea of two systems in the brain, the dopamine-based 'give it to me, I want it now' system, and the more reasonable prefrontal system, which appreciates that sometimes it's better to wait. Some researchers have taken to calling these systems 'now' and 'later'. But as you might suspect, when we look more closely at one of these systems, it turns out it's not that simple. In fact, it seems that different areas of the striatum respond differently to delayed reward. One part of the striatum is involved in inhibiting actions, another is more important for reward and impulsive choices, and a third has a role

* Of course, this isn't the only reason people don't save money-high rents and low wages mean many just don't have anything left over after buying the basics.

in habits. The prefrontal cortex, equally, is not one area with one function. Parts of it help sustain effort, while other areas are more active when someone chooses impulsively. Each of these connects to the corresponding section of the striatum.

By now, this level of complexity should be starting to feel familiar, as we've seen it again and again. But it turns out there is another layer of complexity we haven't yet covered.* Even within the same populations of neurons, the *way* they fire is important, and can signal different things. In fact, 'now' and 'later' might be better referred to as processes, rather than systems. For 'later' to win, a steady level of dopamine must be present in the striatum and prefrontal regions of the brain. This keeps the animal motivated to continue pursuing the goal of the delayed reward. This kind of firing is known as tonic. It can be thought of as a background level of activity, like the gentle classical music played in a fancy wine bar. 'Now' processes, however, are driven by sharp, sudden bursts of dopamine – think a sudden clash of cymbals or a clip of death metal, playing over our background tune. This would, of course, grab your attention, and it does the same in the brain, driving us to want whatever it is that caused it NOW please! This is the 'reward' or 'wanting' signal we discussed in Chapter 3.

I want it NOW!

So it is competition between these two systems that allows us to weigh up our options and decide whether to take the smaller reward now or a larger one later. Dopamine changes in the prefrontal cortex signal how

* I can hear you all cheering!

available rewards are. Less dopamine here means we discount future rewards more dramatically. Meanwhile, the nucleus accumbens tells us about the value of the reward on offer, and more activation here leads to us choosing instant gratification. The ventromedial prefrontal cortex also has a role in assigning the value of a reward, based on how much the animal needs that reward. As we saw in Chapter 6, food tastes better when we are hungry, and this is thanks, in part, to the links between this area of the prefrontal cortex and the nucleus accumbens. And damage to the ventromedial prefrontal cortex (as experienced by EVR and described above) causes an increase in temporal discounting.

There are other regions involved in 'now' or 'later' decisions too. The hippocampus, for example, allows us to use our memories to imagine each outcome, based on past experience, which affects the value we assign to the options. The amygdala gets involved too, bringing emotions into the mix. Just as in the Iowa gambling task, damage here can increase risky choices, with people accepting large rewards now even though they would lead to large losses later. It seems the prefrontal cortex has a role in comparing information from all these different areas, and weighing it to help us (hopefully!) make the best possible decision.

The interconnectivity of these two systems means that if we are distracted, engaging our prefrontal control areas less, the nucleus accumbens becomes more active, resulting in decisions based more on the value of the reward. Similarly, when we try to overcome cravings, like that chocolate bar we fancy but don't need before dinner, we use our prefrontal regions to damp down activity in the nucleus accumbens and control the urge.

Then there are the receptors. As we have seen before, the same neurotransmitter can have different effects based on which receptors the neurons in a certain area are using. In the prefrontal cortex, there are two different dopamine receptors, one of which tips the balance towards riskier options, and the other towards safe choices. Similar differences can be found in the nucleus accumbens. This could explain some of the differences between individuals in terms of risk-taking, as someone with more of the first receptor might be more of a risk-taker. It can also account for how these tendencies can change over the course of your life,* as the number of each receptor can be altered.

These circuits are vital for us to make good choices about the future. It may be issues with their balance that can lead to drug abuse, gambling addiction and overeating, all issues we have seen in previous chapters. Unfortunately, behaviours like these, which provide fast, dramatic rewards, can skew the system even further. If the behaviours are repeated, this can reduce the activity of the frontal regions, make the amygdala and hippocampus more reactive to stress and reward cues, and make people more aware of desire states. All of these make it harder to resist the immediate reward next time, leading to a vicious cycle.

So what can we do to give ourselves the best chance of making the right decisions for the future? Interestingly, putting yourself in the shoes of 'future you' can help.

---

* Generally, we tend to be at our most risk-seeking during our teenage years, becoming more cautious as we enter adulthood, although this varies depending on the type of risk being studied.

Using software to age a picture of your face, for example, and looking at that while making decisions about your future can help you to be more generous towards your future self.

There are other ways we can get around this bias. If you have ever refused to buy junk food when you are on a diet, even though your family would eat it, you have used a commitment device. Others include locks on credit cards (or even freezing them in a block of ice!), promising only to listen to that brilliant podcast when you are out running, or signing up for a charity race. 'Current you' has made a commitment that 'future you' can't break, and that can help you stick to your goals.

These tricks highlight an important point. While neuroscience research is fascinating, and we are starting to understand the networks and chemicals that underlie various aspects of our lives, it is still very much in its infancy. As we develop better techniques that allow us to look at the brain in more detail, our understanding is changing. Things we thought we knew a year ago are being updated daily, and it is almost certain that several 'facts' I have written in this book will be out of date by the time you read it.

This doesn't mean we should give up on neuroscience, but it does mean that when looking at the practical actions we can take to understand ourselves better and improve our lives, we are better off looking at a body of research that has been around longer: that of psychology. Neuroscience might try to tell us why, but it is psychology research that shows that stress makes us worse decision-makers, and how we might be able to reduce our stress levels (see Chapter 4). And it is psychology that allows us to be aware of and counter the biases we might fall foul of when making decisions. We might not yet understand exactly how it does it, but behavioural changes *can* affect our decision-making circuits, and the chemicals that control them, and shift our decision-making capability. And this realisation couldn't really have come at a better time, as we are about to dive into a topic which involves what is, for most people, the biggest decision they will ever have to make.

CHAPTER EIGHT

You've Got the Love

Love and attraction are a huge part of life, for most people. While for some it may be 'all you need' and others think it's 'a four-letter word', it has fascinated musicians, poets, writers and philosophers for thousands of years. But it's only been more recently that scientists have got involved. With their usual love of categorisation, they have broken down the wonders of romantic love into three elements: lust, attraction and attachment. Each of these has a specific pattern of brain activity and neurochemistry involved, although there is a lot of overlap. So, I decided to start at the beginning, looking at probably the simplest of these elements, from a scientific perspective. Lust.

Before we dive into this chapter, I want to start with a caveat. Human sexuality and gender is extremely complex. The more we learn about it, the more we realise that we are all on a spectrum when it comes to how we identify and who we are attracted to. Unfortunately, most of the studies quoted in this chapter haven't really caught up with this. Either the researchers didn't think about including diverse participants, or they decided it was too complicated. This means that the majority of the studies involve straight, cis-gender people. The few that did include homosexual couples, or didn't ask about preferences, grouped them in with the straight couples, or didn't include enough of them to see if there was a meaningful difference.

This makes a bit of sense when we are trying to compare human sexual behaviour with animals', which tends to be between a male and a female (although not always*). But it does mean there are huge swathes of the population who we just don't know much about when it comes to love and attraction. We can't be sure whether these studies extrapolate to people who aren't heterosexual. And we can't be sure how transgender and non-binary people fit into the picture. Or how non-monogamous relationships differ from monogamous. This is a real shame, and I hope that, in the future, more studies will include more diverse participants. While I have tried to be as inclusive as I can with my language throughout this chapter, I hope you can forgive my focus on straight, monogamous relationships. This isn't because I think these are any more important than other types, it's simply because that is where the majority of the studies have been done so far.

Lust for life

Lust is the drive to have sex, something which almost all animals need to do to pass on their genes to the next generation. This means it is a strong urge, driven by

* Bottlenose dolphins, for example, indulge in homosexual behaviour as much as heterosexual, and two male chinstrap penguins in New York City's Central Park Zoo made headlines when they began incubating a rock as if it was an egg. The pair were later given a real egg, taken from a heterosexual pair who were unable to raise it. They successfully incubated the egg, and raised a daughter. For more stories about the complexities of animal sex and mating, I recommend *Sex on Earth* by Jules Howard (Bloomsbury Sigma, 2014).

evolution. But humans are unusual, in that our lust has become disconnected from procreation.

We don't know whether animals feel lustful in the same way we do, but they certainly have the desire to have sex, and will work hard for it. But in most mammals, most of the time, it is tied to fertility. Usually, a female cat (quite understandably) has no desire to have a male's barbed penis anywhere near her, but when she comes into heat, something changes. Suddenly finding a mate is at the forefront of her mind, as her brain is flooded with the reproductive hormone oestrogen. And a male cat, on hearing her seductive yowls, will instantly be aroused and ready to go.

The same applies to mice and rats, the mammals most studied in labs around the world. When a female reaches the fertile part of her cycle, she begins exhibiting strange behaviours. When a male is near, she will hop and dart. If he (or a researcher) touches her on the flanks, she will arch her back in a posture known as lordosis. A female not at peak fertility behaves completely differently around males. The difference? The hormones oestrogen and progesterone in her brain. Lordosis is a reflex. Tactile signals travel to the spinal cord, causing her motor neurons to move her into the receptive position. But most of the time, the reflex is inhibited by part of the hypothalamus. When she is in heat, however, one of the oestrogen hormones changes activity in this area, allowing the reflex free rein.

In humans, lust isn't as closely tied to reproduction. A woman can be turned on at any time of the month (although some studies do suggest she might be more receptive to sexual stimuli around the time of ovulation). And people can be attracted to other people who they aren't able to make a baby with. But our desires are driven by the same hormones: testosterone and oestrogen. This is why our interest in sex begins around puberty, when these hormones flood the brain and body, causing them to change in a whole host of ways. And it is why our sex drive tends to dull as we age, as the production of these chemicals drops. It even explains why someone who is pregnant may find their partner (or favourite sex toy) irresistible at certain points in their pregnancy.

Lust is an important drive, but it isn't necessarily focused on one particular person. So, while it can be an important factor in relationships, particularly at the beginning, there is something more complex we need to look at when thinking about human romance.

Neither my husband Jamie nor I really remember the first time we met, and only partly because of the alcohol we had both consumed that night. I was in my first year at university, he was in his second year; it was Freshers' Week, and we were in the college bar. We both met a lot of new people that night, and the following day all I remembered about our encounter was that I had talked to some second year about salsa dancing. Luckily, when I told a friend this in the bar the following night, Jamie's roommate overheard, and realised who I was talking about.* He (re)introduced us, and we began meeting once a week to dance.

There was no sudden bolt of lightning when we met, no instant desire or lust. In fact, he had been chatting up another girl that night, and I still had a boyfriend from home. But as we got to know each other, we became friends, and the attraction between us grew. So what was

* Apparently there weren't that many keen salsa dancers in his year!

it that drew us to each other, over those weeks and months? How did we go from strangers in a bar, to deciding we wanted to spend the rest of our lives with each other? It all started with a growing attraction.

Across a crowded room…

Attraction is much more specific than lust. This is the part where you decide you like someone, and that they might be more than just a friend. A huge range of brain areas are involved when we see someone we find attractive, particularly parts of the reward circuit and emotional areas like the amygdala. But when we look closer at these studies, once again, things aren't quite so simple. The amygdala, for example, is more active when looking at attractive rather than middling faces, but is also more active when looking at unattractive faces.*

* An interesting issue here is who defines the face as attractive or unattractive. Often this is done by asking a large sample to rate the attractiveness of a set of images, and then averaging the results. But it is easy to see how this can be biased by the sample. If the raters are all university undergraduates, for example, as is so often the case, they will be much more similar than a random sample of the population would be. You might get very different results if you sampled a wider cross-section of people. And the ratings will definitely be influenced by culture, so extrapolating this as an objective measure of attractiveness is a jump. And sometimes the issue is more blatant. In a recent study, nearly 200 university students were asked how attractive they thought they were, and these ratings were compared to their 'objective' attractiveness, to see how realistically they saw themselves. The measure of 'objective attractiveness'? 'While participants responded to the questionnaire, their objective attractiveness was unobtrusively assessed by two male experimenters who were standing in front of the participants while they were filling out the questionnaire.' Need I say more?

But, again, the majority of these studies are done using young, heterosexual men, looking at pictures of women. In fact, the few studies that have involved women as subjects found different patterns of activation.

When it comes to the reward system, things are also not clear cut. One study found activation in parts of the reward system only if the attractive face was looking directly at the subject. For faces looking away, the more attractive they were, the less active part of the striatum was (the striatum, as we have seen, is involved in the reward system and decision-making). This makes sense. Making eye contact with someone you fancy is a good thing, and your reward system might release dopamine to help you to learn to continue what you are doing, and to approach the person. If they pointedly avoid catching your eye, a dip in dopamine would teach your brain that what you are doing isn't working.

So, does this tell us that dopamine is the brain chemical involved in attraction? Not quite. When male mice were given drugs to boost their dopamine levels, it didn't change the rate at which they approached sexually receptive females, and blocking dopamine didn't reduce it. The same was found in females. In fact, opioids in the reward system might be more important. In a study led by Olga Chelnokova at the University of Oslo, giving human volunteers morphine led them to rate attractive faces as more attractive. They also looked longer at these faces, and pressed a button more times to keep them on the screen. Blocking the opioid system had the opposite effect.

But the attractiveness boost only happens for the most attractive faces. Manipulating this system didn't make intermediate or less attractive faces seem prettier. In fact, participants moved on from the least attractive faces more quickly after being given morphine. These findings suggest the opioid system is involved in both 'wanting' and 'liking' attractive faces (see Chapter 3 for more about wanting versus liking), but in a complex way.

As we can see, the processes that occur in the brain when we find someone attractive are still poorly understood. This isn't really surprising, considering attraction is probably the part of love that seems most mysterious. Why does one potential partner set your heart racing and your stomach a-flutter, while the idea of another leaves you cold? We've still got a long way to go, but science is starting to find the answers.

Of course, what someone looks like plays a big role in whether you are attracted to them. In cultures all around the world, there are certain traits that are deemed attractive, such as symmetrical features, certain facial proportions and clear, bright skin and eyes. Some evolutionary psychologists believe these traits act as markers of genetic fitness. Anything non-symmetrical, the theory goes, may be down to a problem in the developing embryo, for example. Picking a non-symmetrical partner might, therefore, put your offspring at risk for inheriting whatever genetic defects they have and that might make them less likely to survive. So, over time, those who picked the healthy symmetrical partners were more likely to have surviving offspring, and we evolved a preference through a mechanism known as sexual selection.

This mechanism has been suggested to explain our other preferences too. Heterosexual Western men, for example, tend to be attracted to women with a youthful appearance, full breasts and waists narrower than their hips, with the 'perfect' waist-to-hip ratio often touted as 0.7.* These are all argued to be signals of fertility, in some cases through a link to hormones. Oestrogen encourages fat to be deposited on the thighs and buttocks, while testosterone pushes it towards being stored around the middle. We see this most clearly around menopause: when women's oestrogen levels drop, they tend to lose their defined waists, at least to some extent. So, the argument goes, a woman with a small waist is likely to be more fertile, because of her hormone balance.

There are, however, problems with the original waist-to-hip ratio studies. Many of them didn't consider overall apparent weight when manipulating the images to change the waist-to-hip ratio, and it turns out that BMI is a bigger determinant of attractiveness than waist-to-hip ratio. And a study that included even more dramatic ratios, down to 0.5 (such as images with measurements of 40-20-40 inches, a ratio similar to a Barbie-doll), found men preferred these, even though they are naturally unobtainable. This is hard to defend from an evolutionary point of view.

Cross-cultural studies have also shown that these preferences aren't constant in all societies. In fact, it seems that higher testosterone levels (and the larger waists they bring) might be beneficial for women in some environments. This throws doubt on the theory that there is a universal, innate body shape preference. If this preference had evolved because it provided a big

* Marilyn Monroe, with measurements in 1945 of 36-24-34 inches, comes out with the 'perfect' waist-to-hip ratio of 0.7.

advantage and increased the chance of successfully passing on your genes, it should be present everywhere. Instead, it suggests an environmental factor.

We clearly do have preferences when it comes to how we want our partners to look, but perhaps these are more culturally than evolutionarily programmed. And there are also individual preferences. Some people, for example, find bodybuilders irresistible, while I personally tend to be much more attracted to lean physiques rather than those that look like they were honed by hours in the gym. This is likely to be, at least in part, because of my experiences and the culture I have grown up in. But what exactly is going on in my brain to cause my preferences isn't yet well understood. And, of course, it isn't just what someone looks like that determines whether we are attracted to them.

Darling, you smell lovely

In fact, there are more subtle cues that can lead to attraction, and an important one is smell. We know that humans can detect a lot of information from one another's musk. We can recognise our family members from their odour, and even determine facts about strangers such as age or, amazingly, personality.*

In many animals, there is evidence that individuals choose partners on the basis of differences in their

---

* One study found people could judge others' extraversion, neuroticism and dominance about as accurately by sniffing a smelly T-shirt as previous studies had achieved through watching their behaviour on video. Interestingly though, ratings of dominance were only accurate when people rated the smell of a member of the opposite gender.

immune system. The major histocompatibility complex (MHC) is a set of genes involved in detecting pathogens and activating the immune system, and it is passed on from parents to their offspring. To have the healthiest offspring, it may be an advantage to have parents with *different* genes in their MHC. Doing so would also avoid inbreeding, as those with MHCs similar to your own are more likely to be relatives. And it seems some animals can detect differences in potential mates' MHC via smell. Each individual is constantly releasing tiny fragments of protein, bound to their MHC, via bodily fluids. These are broken down by bacteria on the skin, to give each animal their characteristic smell. And this, the theory goes, guides mate selection.

But does it work in humans? There was a lot of excitement for the idea in the 1990s when Claus Wedekind, now Professor of Biology at the University of Lausanne, Switzerland, carried out experiments on – as usual – college students. He gave T-shirts to a group of male students, and asked them to wear them for two nights, while avoiding the use of deodorants or scented products. He then presented these T-shirts to a group of women, and asked them to rate how attractive each scent was. Analysing the results, the team found that the women consistently chose men with MHCs different from their own.

So does that mean we have no control over who we end up with, and pick them solely based on their stink? Of course not. In fact, there have been mixed results when other groups have tried to replicate these findings, with some finding no effect at all, so we can't treat them as decisive, even in the lab. And even if it were the case that humans prefer the body odour of others with dissimilar immune systems, or perhaps dislike those with very similar genes, this doesn't tell us whether we actually *use* this as a factor in choosing a partner. Certainly, if you ask someone why they are with their other half, they are likely to talk about their intelligence, humour, kindness and potentially their appearance, but are unlikely to mention their natural musk. And most of us try to prevent body odour as much as possible. Washing regularly and using scented products is generally seen as a way to attract a mate,* not put one off.

To work out whether MHC differences really do affect mate choice in real-world environments, a number of studies have taken couples, sequenced their genomes and compared them to random pairs of individuals from the same population. Again, results are mixed. One study, led by Sara Pulit from the University Medical Center Utrecht in the Netherlands, found partners were no more genetically dissimilar than strangers. Another, led by Raphaëlle Chaix at the University of Oxford, on American subjects of European descent, found there was a difference. A re-analysis of Chaix's data, however, showed that this could be down to just a few outliers in the non-couple group. So it seems the jury is still out on whether sniffing someone's genes can make you want to get into their jeans.†

Despite these unknowns, it does seem that scent is an important part of attraction and desire, both romantic

* Or a whole swarm of potential mates, if Lynx adverts are to be believed… and they definitely were believed by all my male friends when I was a teenager!

† Sorry, couldn't resist!

and sexual. And this becomes clear when we look at people who don't have this sense. Anosmia (the inability to detect smells) can be something you are born with, but more commonly it occurs later in life, either because of an infection, or because of head trauma, which can damage or even sever the nerves leading from the nose to the brain. There is evidence anosmia affects social relationships. A series of studies led by Ilona Croy at the University of Dresden, Germany found that people born with anosmia are more socially insecure than healthy controls, and that men with the condition typically have fewer sexual relationships than average, while women feel less secure in their relationships. About a quarter of those who develop anosmia report reduced sexual desire, and many say it affects their enjoyment of sex. Patients are also more likely to be depressed.*

This makes a lot of sense based on my experiences. I am sure I am not the only one who likes to wear their partner's jumper when he is away, and I definitely get a sense of comfort from his unique smell. What isn't clear, however, is whether it is something specific in his scent that makes me feel like that, or if I have simply learnt to associate it with him, and therefore with feelings of comfort, safety and attraction.

So it seems that smell is related, in some way at least, to attraction. But one word that is often thrown around when talking about these studies is 'pheromones'. In fact,

---

* This fact does introduce a complicating factor for these studies, however. If anosmia leads to depression, it may be the depression that has an impact on relationship formation, rather than the anosmia itself.

you can buy bottled pheromones from websites around the world, claiming to make you irresistible to men or women. And studies like the smelly T-shirt ones are often touted as evidence that pheromones exist in humans, and that they are involved in attraction. The reality is much less clear cut.

A pheromone is simply a chemical messenger that transmits signals between individuals of the same species. They exist in fish, insects and mammals and can be transmitted long distances, like the sex pheromones that attract male butterflies to females up to 10 kilometres (more than six miles) away. Others are exchanged in intimate encounters, like the substance produced by a queen honeybee, which prevents her workers' ovaries from developing. They can carry all sorts of different types of message, but often they are involved in mating.

There are records back as far as the seventeenth century giving examples of long-distance communication amongst animals, and Darwin talked about the 'musky odour' of the male crocodile. In the 1800s, French entomologist Jean-Henri Fabre carried out a series of experiments on giant emperor moths, proving that their sense of smell helped them find a mate. He found he could prevent the male from finding the female by putting her in a closed box, or by removing the male's antenna. He also discovered that the male would be attracted to a cage the female had previously occupied. He may not have known it, but he had discovered pheromones. Despite this early discovery, it took until 1959 for the first pheromone to be identified.

Adolf Butenandt was a German chemist, who won the Nobel Prize in 1939 for his role in identifying sex hormones oestrone (one of the three major oestrogen hormones), androsterone (an androgen, the class of hormones that includes testosterone) and progesterone (a hormone involved in the menstrual cycle and pregnancy). His next aim was to work out what molecule it was that attracted male silk moths to females. His study subject was judiciously chosen, as the silk industry in Europe was booming at that time, so he could easily get hold of plenty of the creatures to investigate. Butenandt and his team extracted chemicals from the glands of the female moths, and set about analysing them.

This proved to be a huge challenge. The chemicals they were interested in are produced in tiny amounts, and the team needed to separate them from all the other molecules extracted. So they developed a technique called a bioassay, using the male moths to examine each solution, waiting for the characteristic wing-fluttering response that signalled his desire to mate. They continued purifying solutions, aiming to find a material that could produce this 'flutter dance' at the lowest possible concentration.

It took over 15 years, but eventually, the team found what they were looking for, and set about identifying the molecule. A year later they succeeded, and in 1959 they confirmed its identity by synthesising it from scratch. They called the silk moth molecule bombykol, and the class of signals pheromones, from the Greek for 'transferred excitement'.

We've got chemistry

This finding is all very well, but we are not moths. So what is the evidence that humans have pheromones? It turns out, there is surprisingly little. Though often quoted in support of the theory, the T-shirt studies are actually looking at individual smell, not pheromones. Pheromones, by definition, aren't individual. For them to work as a signal between different members of a species, they have to work on everyone. Any male moth sensing bombykol will find it attractive. That means if one of the men in the T-shirt study was producing the 'irresistibility pheromone' so loved by advertisers, *every* heterosexual woman in the study should have preferred his scent. But this isn't what was found. Time and time again it seems to be personal preferences at play.

Then there is the question of whether humans can even detect pheromones. In most animals, there are two olfactory systems: the main olfactory system, which detects scent, and the vomeronasal organ, which detects pheromones and uses them to trigger behaviour. If the vomeronasal organ is damaged, male mice won't show interest in a fertile female's urine (although they will still mate with her if she is present), and females won't display lordosis. Male hamsters will mount other males if they are sprayed with the pheromones usually produced by females in heat.

There is debate over whether humans have this vomeronasal organ, and, if we do, whether it is still functional. It does seem we once had one, in evolutionary history. In fact, it even begins to develop in embryos, but then regresses. Studies have found remnants are present in some adults, but it doesn't seem to be functional; there are very few nerve cells, and those that have been found aren't connected up, meaning it has no way of detecting pheromones or sending signals to the brain to control behaviour. And none of the cells express proteins that are markers of olfactory function. Instead, it seems to be made of skin cells, so is very unlikely to be usable.

But even without a functional vomeronasal organ, we *could* still detect pheromones using our main olfactory system. Pheromones could diffuse into the blood through capillaries in our nose, or directly travel into the fluid around our brains. Or they could activate receptors in the main olfactory system, sending signals to the brain to change our behaviour. But evidence for any of these methods is lacking.

Despite this, there are widely published claims that four molecules are human pheromones: androstenone, androstenol, androstadienone and estratetraenol. And these are the molecules that tend to be found in the pheromone sprays that can be bought online. So where did this idea come from? The first two of these are indeed pheromones… at least in pigs. So, when scientists discovered tiny amounts of these compounds in human armpits, they jumped to the conclusion that they must be pheromones for us as well. But evidence to support this conclusion just doesn't stack up. As Tristram Wyatt, Senior Research Fellow at the University of Oxford, wrote in his 2015 paper on the topic:

> What likely made androstenone popular with experimenters was the commercial availability of the molecule in aerosol cans as Boarmate™, for use in pig husbandry… The more fundamental criticism is simply that there was no evidence to justify using these molecules with humans in the first place, rather than any of the hundreds of other molecules found in human armpits.

The second pair, androstadienone and estratetraenol, known for short as AND and EST, have a slightly shadier history. In 1991, a conference was held, with scientists presenting their work, as tends to happen at conferences. But this event was sponsored by a company called the Erox Corporation, which had patented several 'putative' human pheromones. The word 'putative' is important here, meaning 'believed to be'. So, the company didn't have to prove these were pheromones to get their patent. And neither did they provide any proof when they presented their papers at the conference. As Wyatt writes:

> The authors gave no details at all of how these molecules were extracted, identified, bioassayed and demonstrated to be the 'putative pheromones' of the paper's title, just a footnote: 'These putative pheromones were supplied by EROX Corporation.'

The molecules were identified in the patent as AND and EST, but only really picked up significant scientific interest thanks to a paper by Suma Jacob and Martha McClintock, published in 2000. McClintock, at the University of Chicago, had made a name for herself with a paper claiming that the menstrual cycles of women living in close proximity 'synced up', and hypothesising, based on research in mice, that this might be to do with pheromones. This concept caught the public's attention, and is still referenced in popular culture despite the lack of scientific support (a recent review paper found no evidence for it). So, when she published another paper on pheromones, using AND and EST, it caused a stir.

And Wyatt believes this is where all the more recent pheromone research stems from:

> Jacob & McClintock has been cited more than 150 times (Google Scholar, 23 October 2014) and the molecules and the concentrations used by them have become traditional across much of the field to this day. However, while Jacob & McClintock were fairly cautious, commenting that '… It is premature to call these steroids human pheromones', this advice has been forgotten by most later authors.

And so we end up in a situation where the pheromones used in almost all pheromone research might not actually be pheromones. But then why are people finding that these molecules affect human behaviour? Wyatt argues publication bias may be to blame.

Publication bias is probably the biggest issue facing science today. To do research, scientists must get funding. And funders tend to fund scientists who have published research in top-tier journals, like *Science* and *Nature*. These journals like the juicy stories, with splashy results and dramatic take-away messages. 'Pheromones change human behaviour' is likely to get published. 'Pheromones have no effect' is less likely to grab the headlines.

This means scientists often don't even try to get their negative results published, something known as the 'file-drawer effect'. We can't know how many pheromone studies have been conducted which found no effect. There is also the issue that generally funders don't like funding replications of studies (again, because they are less likely to be published), so while replication is the cornerstone of good science, scientists often don't get the chance to attempt to replicate each other's work.*

So where are we with the understanding of human pheromones? Sadly, Wyatt argues it is time to go back to the drawing board. 'If we are to find pheromones we need to treat ourselves just as if we were a newly discovered mammal. We need a scientific and systematic search for potential molecules, common to one or both sexes, which have reliable effects.'

The difficulty comes in measuring behaviour, particularly in relation to desire. Humans are incredibly complex, and can be affected by a whole range of inputs, including our culture and experiences. So to find our first definitive human pheromone, perhaps we should look elsewhere – to babies.

Pheromones aren't only involved in attraction, they can carry all sorts of other signals too. People who are breastfeeding, for example, give out a secretion from glands around their nipples which, when placed below a newborn's nose, causes them to start sucking and searching for a nipple. And those who have more of these glands seem to have an easier time feeding their infants. We don't know for sure that this is a pheromone, but it seems promising. Investigating this might not only provide confirmation that we do have and respond to

* The good news is that scientists and journals are becoming aware of this problem and working to address it. One of the ways they do this is by pre-registering studies online, before they are carried out, so that they appear on a list. This means other scientists know they are happening and can follow the results as they come out. Some journals have also decided to commit to publishing the results of studies before they have happened, whatever the outcome.

pheromones, but also help the high proportion of babies who have a hard time feeding, particularly in the first few hours after birth.

A girl walks into a lab...

While there seem to be major problems with pheromone studies, there are also issues with other types of experiment that claim to have discovered the 'key to attraction' in the lab. The biggest one is that most of us (at least) don't fall for our partners in the lab. We don't decide whether to date them based solely on their smell, or fantasise about taking them to bed simply because of their square jawline or low waist-to-hip ratio.

Scientists are careful to control every other element in their studies, making sure each person is viewed (or sniffed) with all other factors taken out of the equation. But this just isn't realistic. I have certainly experienced finding someone attractive initially, but quickly revised my assessment on talking to them. Equally, it is possible to realise you have become attracted to someone who was originally just a friend, based on their company and personality, as happened with my husband and me. Then there is the influence of society. We have it drummed into us from an early age what an 'attractive' person looks and behaves like, so even if we do have preferences, how can we know whether they are really our own, or down to this constant societal pressure and conditioning?

I wanted to find someone who could answer these questions and put the lab studies on attraction into perspective, so I talked to social psychologist Viren Swami from Anglia Ruskin University, UK. He believes that the evolutionary basis for physical attractiveness has been overemphasised. Based on ideas first developed by feminist scholars such as Judith Butler, and Sandra Bartky, and his own cross-cultural studies, he has come to the conclusion that society has a huge role to play in what is seen as attractive. As he explained to me:

> Most cultures propose and promote ideal forms of attractiveness. My view is that a lot of this has to do with patriarchal structures and finding ways of telling people that they're not good enough, in this case mainly women. Society tells us what is attractive, and often these attractive ideals are impossible to achieve for most people. The point is that advertisers have something to gain by making us all feel bad about ourselves.

He agrees there is an important role for physical attraction in the formation of relationships, but doesn't think this is necessarily because we have evolved to pick these 'healthier' attractive people because they are better reproductive choices. For one thing, when sexual selection is at play, we usually end up with dramatically exaggerated characteristics, like the peacock's tail. If women with lower waist-to-hip ratios really were more fertile and found it easier to find mates, we would expect this characteristic to rapidly spread through the population – and it hasn't. In fact, there isn't even any evidence our ancestors cared about attraction. As long as they found a partner who was good enough to mate with, to produce and help raise the children, what they looked like may not have mattered at all! So instead, he argues that our culture is the most important factor in shaping what we find attractive.

And what about smell? Here too, culture is playing a huge role, as Swami explained:

> The problem with applying it to human beings is twofold. Firstly, a lot of the studies are based in the lab where you can exclude other things that are happening. The bigger problem is in most settings we are socially taught that bad things or bad people smell bad, good people smell good. If you turned up to a first date smelling like you came from the gym, you're going to have a bad time.

Does this mean we should throw out all the lab-based work on attraction, and what's going on in the brain when we fancy someone? Swami doesn't think so:

> I think there is value; I think it tells us certain things about what is happening in society at a particular point in time. My main problem is the assumption that because you find someone attractive therefore you will end up mating with that person, or want to mate with that person…

He makes the point that there are lots of people we find attractive, but don't end up having sex with. This can be because they are someone you will never meet (as anyone who has had an intense teenage crush on a pop star will know!); because they, or you, are in a relationship; because they are a flat-earther, or otherwise have beliefs that are too opposed to your own; or even because you believe them so attractive as to be unobtainable. Although your brain chemicals have an important role to play in determining who you are attracted to, it seems this is far from a pre-programmed desire, laid down by evolution generations ago. Instead, it is shaped by our culture and our experiences. And, of course, in the complex world of human social relationships, being attracted to someone is only one piece of the puzzle when it comes to forming a relationship.

Swami explained that social psychologists have identified three other factors important for relationship formation – factors that are so simple they are often overlooked. The first is proximity: we tend to fall for people who are close by. The second is reciprocity: we like people who like us back. And the third is similarity. While it may be their sparkling eyes and cheeky dimples that get you to initiate the conversation, it is your shared interests and, particularly, values that will have you coming back for more. And someone you like and get on with, and who is kind to you, can grow to seem more physically attractive as you get to know them.

These three factors certainly fit with my experience; seeing Jamie regularly, over the course of my first year of university, and learning more about him as we stayed up late chatting about our hopes and dreams, certainly felt like the key to our growing relationship. And while we had wildly different childhoods (he grew up in East Africa, then went to boarding school in the UK when his parents moved to Japan, while I lived in the same area my whole life, and had never even been to Africa or Asia when we met) we had a lot in common when it came to what was important to us. And that, I believe, is what helped us make the move to become more than just friends.

Addicted to love?

However it happens, in some cases attraction leads to a relationship, and after some time, falling in love. And it's that first heady year of being a couple where we see some of the biggest changes in brain chemicals. In early romantic love, people are stressed. This might sound odd, as love isn't usually thought of as stressful, but common experiences like a dry mouth, sweaty palms and a faster heart rate when you see the object of your affection are all caused by noradrenaline. And studies have shown that levels of cortisol are higher in new couples than singles or those who have been together for a long time.

At the same time, the brain's reward system is highly active, producing large amounts of dopamine when you see the person you love. An MRI study found that when looking at pictures of a loved one, compared to an acquaintance, two brain areas in particular were more active: part of the striatum and the ventral tegmental area, which are both associated with motivation and reward. Because of the links between these areas and drugs of abuse, many people have made claims that the feeling of being in love is similar to the 'high' of drugs, but as we saw in Chapter 3, dopamine is not a pleasure chemical, so I think these claims are overstatements. But dopamine does drive you to pursue a goal, so this activity could be responsible for that initial feeling of wanting to be with your partner all the time, and wanting to be close to them whenever possible.

Brain scans have also shown areas that are *less* active when we think about the person we are in love with. These include the amygdala, which is involved in fear, and parts of the frontal cortex, which are important for making critical assessments of other people. This might explain why 'love is blind', and people often fall for individuals their friends can see aren't right for them at all.

Meanwhile, serotonin levels fall. Low serotonin is related to high stress chemicals, so it's likely these differences are linked. Interestingly, serotonin levels are also lowered in obsessive-compulsive disorder (OCD), and some researchers have suggested there are similarities between the early stages of love and this disorder. New couples often think about each other obsessively, and experience stress and anxiety. Then there is the old trope about not being able to eat or sleep because thoughts of the object of your desire are taking over your every moment. Could that be a manifestation of the same types of symptoms people with OCD experience?

There are obviously differences here. The feeling of lovesickness doesn't last long, and in most people it doesn't actually interfere with their ability to live a normal life, unlike OCD, which can be debilitating. But it is an interesting comparison to make, nonetheless. And differences in the gene that codes for one of the serotonin receptors have been linked to obsessive romantic behaviour, supporting the idea that serotonin is involved in the obsessions of early love.

So far all these findings have applied equally to men and women in love, but there are some chemicals that differ between the sexes. In men, testosterone levels drop at the beginning of a relationship. The level of testosterone in women is more ambiguous, as some studies have found decreases for them too, while others found elevated levels. One study found a drop, but only if the partners were in the same city. Women in a long-distance relationship didn't show the same effect. This suggests that, in women at least, it might be the experience of having a partner physically around that is responsible for the change in hormone levels.

Those heady first weeks and months of a relationship are wonderful. I remember well being consumed with thoughts of Jamie, and feeling a thrill of anticipation every time I saw his name pop up on my phone, or his handwriting on a letter.* But over time the initial obsession morphs into something different, though no less wonderful. At this stage, the relationship has entered the long-term bonding phase, where it can remain for a long time, in some cases for the rest of the couple's lives. To understand this phase, and the brain chemistry that drives it, we need to start in what might seem like an unlikely place: rural France in the Middle Ages.

Medicine or poison?

Something strange was going on in the villages of southern France, and many similar villages across central Europe. One after another, people were developing unusual symptoms, the cause of which was completely unknown. Some could be found delirious, suffering from fits and muscle spasms. Others were afflicted with excruciating pain in their extremities. This symptom gave rise to the name given to the mysterious illness: holy fire.†

---

* Just weeks after we started going out, Jamie headed off to spend his summer in the cloud forests of Ecuador, searching for new species of butterfly. This meant three months with minimal contact, but lots of love letters, which I received in bulk when he reached a town with a post office!

† The name was later changed to St Anthony's fire, after the order of monks who were successful at treating the symptoms.

In the late 1600s, with the disease having been prevalent for centuries, a French physician called Dr Thuillier noticed that the pattern of cases of this illness differed from other infectious diseases. It didn't seem to be contagious, as often only one member of a family would fall ill, and it was more common in the countryside than in the packed and filthy towns. And it only struck the poor. Rich people never seemed to be afflicted. This led him to realise that it must be caused by something in the environment.

The culprit, it turned out, was the rye that kept the villages in bread and beer. The grain stores in the affected villages had become infected with a fungus called ergot, which produced a range of compounds that caused the diverse symptoms the villagers were experiencing. And they weren't alone. Some historians believe we have evidence of cases of ergot poisoning dating back as far as the Ancient Greeks, and there are even suggestions that ergot might have been to blame for the Salem witch trials, as the symptoms of 'bewitchment' have many similarities with those of ergot poisoning.*

But as well as producing these dramatic symptoms of poisoning, ergot had long been used as a medicine. First exploited by alchemists and then by midwives, it was used to hasten childbirth. But the compounds produced by this fungus weren't known, so it was a dangerous practice.

Fast-forward to 1904 and we find a young Henry Dale working in the Burroughs Wellcome research

---

* This idea was first put forward by Linnda Caporael in 1976, but has since been the subject of much debate, and there is still disagreement over whether it can explain the phenomenon fully.

laboratory, London. He had been tasked with finding the compound within ergot which produced contractions of the uterus, the hope being that by isolating the compound a safer and more effective medicine could be produced to help with childbirth. Despite not being enthusiastic about the idea, Dale set about conducting experiments, along with chemist George Barger.

Together they successfully isolated a compound known as ergotoxine from the mix of chemicals produced by the fungus, but sadly they found it to be less effective than the ergot mix as a whole, while also producing severe side effects. Continuing his work, Dale then discovered a substance in the mix which acted to dilate blood vessels, and which he later identified as acetylcholine, one of the most important molecules in the nervous system. It was this discovery, alongside others, that earned him the Nobel Prize in 1936.

But it was a finding that is little more than a side note in his study on ergot that is vital for our story. While looking at how ergot interacted with various other substances, Dale noticed that an extract from the pituitary gland of an ox could make a pregnant cat's uterus contract. He called the substance oxytocin, from the Greek words for 'swift birth'.

His discovery was only recorded as a throwaway comment, with the real focus of his work being on ergot and its effects, but it is an important one. Not long after, it was discovered that injections of the same pituitary extract caused an increase in the amount of milk produced by a lactating goat. But it wasn't until the 1940s that this work was confirmed, and it took decades for the hormone involved to be isolated and characterised.

In the early 1920s, Vincent du Vigneaud was studying biochemistry at the University of Illinois when his interest was sparked by a lecture about the discovery of insulin. This set du Vigneaud on a career studying insulin, which quickly led to him investigating the pituitary extract Dale had been working on. He managed to identify two hormones in the extract: vasopressin and oxytocin. It took until 1953 (with a brief hiatus to work on penicillin during the Second World War) for du Vigneaud to first synthesise oxytocin in the lab. This work led to him winning the Nobel Prize in 1955.

Its synthesis opened oxytocin up to the medical world. Testing found that the synthetic version was indistinguishable from natural oxytocin and could bring on labour at full-term, as well as induce milk ejection in someone who is nursing. This proved that oxytocin has an important role to play in birth and lactation. We now know oxytocin levels rise at the beginning of labour, causing the contractions of the uterus that push the baby out. After the birth, oxytocin levels remain high, helping to deliver the placenta and contract the uterus to reduce bleeding. During breastfeeding, oxytocin helps to eject milk and is released when the infant suckles. But could the hormone be doing something more than producing these simple physiological responses? To answer that question, first we need to get acquainted with one couple and their goats.

The parent switch

When a kid is born (I'm talking about goat kids here, not human kids), its mother will look after it, and allow it to suckle. But if another goat's kid tries to suckle from her, she will push it away.[*] In the 1960s and 1970s, scientists wanted to find out how this worked. How does a goat recognise her own offspring, and know to look after it, while rejecting others? They soon realised this was actually two distinct questions. Smell appeared to be important for recognition, but maternal bonds could still form when this sense was impaired.

In a series of experiments, Peter and Martha Klopfer, working at Duke University, US, set about taking newborn kids away from their mothers and returning them after different lengths of time. In some cases, they replaced the kid with another one, from a different mother. They found that as long as the mother had a kid to lick directly after giving birth, whether her own or another, her maternal instincts would kick in, and she would later accept her own kids. But if the new mother had no contact with a kid immediately after giving birth, she would reject even her own offspring if they were returned to her just an hour later.[†]

In their 1967 paper, the Klopfers speculated as to why this might be. They noted that oxytocin is known to spike around birth and fall to normal levels within a few minutes after. 'It is tempting to speculate that this hormone, which apparently brings on the final uterine spasms which deliver the kid, is also implicated in the induction of maternal behavior.'

---

[*] This often happens – kids are fairly indiscriminate and will even try to feed from an animal of a different species.

[†] Interestingly, a mother who had licked another animal's kid straight after the birth would also accept that kid if it was taken away and then returned to her, but not other kids belonging to strangers.

Experiments have also shown that oxytocin is important for maternal behaviour in rodents. Females that have never given birth are fearful of pups, and will avoid them whenever possible. But as soon as they give birth, within a couple of hours of the birth their behaviour switches. Instead of staying away from the young, they will care for them and defend them with their life. They will even, in many cases, accept others' pups as if they were their own.

This switch is, again, down to hormones. Inject oxytocin into the brains of virgin animals at the right point in their cycle and they begin to show maternal behaviours. Block it in a pregnant female and she won't build a nest, or care for her young once they are born.*

So how does this change happen? Bianca Marlin at Columbia University discovered, in a study on mice, that there are oxytocin receptors in the auditory cortex, and when the hormone binds here it changes the way the area responds to the pup's calls. It may be that it has similar effects in other brain regions involved in caring for offspring.

In humans too, there is evidence oxytocin is important for the parent–child bond. Studies have shown that parents with higher levels of the hormone have stronger bonds with their infants, and giving a parent oxytocin can boost activity in the parts of their brain involved in empathy and reward. This might drive the desire to

---

* In some cases, however, virgin mice *will* spontaneously begin to care for pups they are presented with after being around them for just 15–20 minutes. Rats can exhibit this behaviour too, but it takes about a week of exposure to the pups to kick in.

interact with their child, through those same pathways that drive us to eat or take drugs. But there is also a feedback loop here, as interacting with a child boosts oxytocin, allowing parents who haven't given birth to bond with their infants.

After the Klopfers' work, the evidence rapidly stacked up that oxytocin was vital for the love between parents and their children, but it took another discovery in animals for scientists to realise that it was also important for romantic relationships – a discovery that involved some extremely devoted voles.

Voles in love

In 1971, Lowell Getz was Professor of Zoology at the University of Illinois and had just launched a 25-year study of the populations of rodents in the fields around the university. To investigate this, he and his team would lure the animals into traps before recording, marking and releasing them. But as they sat in the draughty research shed for hours on end, they began to notice something odd. Some of the voles were repeatedly appearing in pairs, usually one male and one female. Not only that, but they seemed to stick with their partner for months, as the researchers would sometimes capture the same pair multiple times. This was surprising, as fewer than 5 per cent of mammals are monogamous. But it seemed it was only one species of vole living this coupled-up life: the prairie vole. Other closely related species were almost never caught as a duo.

To find out more, Getz chose 12 of the prairie voles to be fitted with tiny radio trackers on collars, so he could follow their movements. Amazingly, 11 of the 12 couples lived together in underground dens and seemed to be permanently partnered up. And the final couple? It turned out that they each had a partner waiting at home for them and were captured while 'playing the field'!

Getz took these findings to a colleague at the University of Illinois who was investigating hormones in hamsters, Sue Carter. Carter was amazed by his discovery, having never encountered bonding like this in rodents.* Together, the pair studied the bonding of these animals, and discovered that attachments formed instantly. As soon as a mature female sniffed the urine of a non-related male, she was hooked. The pair would enter into a marathon mating session, which could last for up to 40 hours, and that was it, they were bonded for life.

This finding sparked the beginning of decades of research into the neuroscience of monogamy, starring these furry critters. Prairie voles are great for research as they can be compared to the closely related montane voles, which aren't monogamous. There are other differences between these two species as well. Prairie vole parents take care of their offspring together, and for a long time (by rodent standards). In montane voles, parental care is minimal, and entirely the responsibility of the mother. Prairie voles are social, spending a lot of time with their partner, even when not mating or caring for young, while montane voles prefer to live a solitary life.

---

* In an interview for *Smithsonian* magazine Carter mentions her surprise at this finding, explaining that she was more used to working with female hamsters, which, cuddly as they may seem, often kill and eat their partners after sex.

The next big question was why. What differed between these two animals to cause this huge difference in behaviour? Perhaps unsurprisingly in the context of this chapter, the answer turned out to be, at least in part, oxytocin. Carter and her colleagues took female prairie voles and gave them a choice. They could stay in the room they started in, alone, or go through one of two doors. Their partner was in room 2, while in room 3, they would find a male they had never met. As expected, the female preferred her mate, choosing the box with him in it significantly more than the new male or her original box. Then, the experimenters began exploring what needed to happen to create this mate-preference effect. They found that if the couple had been living together for 24 hours, even if they weren't allowed to mate during that time, the preference would form. But it could also form more quickly, in just six hours, if the pair were given the opportunity to have sex.

But what was interesting was that the researchers could recreate this difference by injecting oxytocin. If a female spent six hours with a partner, and didn't mate with him, but was given oxytocin, she would choose to spend most of her time with that partner during the test. And other studies showed that if you blocked oxytocin receptors, a female who had recently mated wouldn't show as strong a preference.

This suggests mating causes an increase in oxytocin, strengthening the pair bond, and other studies have shown that this is the case. Researchers think this might be a case of a hormone for one type of behaviour (maternal care and parent–child bonding) being co-opted into another (partner bonding), something that is very common in evolution. The marathon mating sessions of prairie voles means the female experiences a lot of vaginal and cervical stimulation, which might release oxytocin in the same way childbirth does.

But what about the males? Just like females, recently mated males show a strong preference for their partner over a stranger. Studies disagree on whether this is influenced by oxytocin, with some finding it does have an impact on bonding, and others finding it doesn't. But there is another very similar chemical called vasopressin, which has a more definite impact on bonding in males.* Just like the females given oxytocin, male prairie voles given vasopressin will become bonded to a partner they have spent just a few hours with, without mating. And blocking the chemical prevents them forming a bond with a female they have lived and mated with. Interestingly, once they have mated, male prairie voles become aggressive towards other males. The once chilled-out animals will attack any new males that come near him or his mate, something that doesn't happen with montane voles. Again, vasopressin seems to be responsible.

Comparing prairie voles to montane voles, a striking difference was found. Giving the montane loners a dose of oxytocin or vasopressin didn't turn them monogamous, as you might expect. Instead it had very little effect at all. So why would the same chemical

---

* It seems vasopressin might also be involved, to some extent, in female bonding. Many researchers now believe both chemicals affect both sexes, but it does seem females may be more responsive to oxytocin, and males to vasopressin.

cause such different reactions in two such similar species? A clue came from looking at their genomes. The two animals have DNA that is 99 per cent identical, but that 1 per cent difference is in genes relating to the way their hormones work. And it turns out the important difference is where in the brain the receptors for oxytocin and vasopressin are found.

Prairie voles have receptors for these molecules in the reward circuits of the brain (see Chapter 3). Triggering these areas when they are with their partners means they learn to associate them with the rewarding effects of sex and that drives them to spend more time together. Block these hormones, and sex becomes short lived, more like the montane voles, and you lose the bonding effect.

But even within prairie voles, genetic differences have important knock-on effects. Remember that pair of voles Getz found that were 'playing away' while their partners waited at home? This isn't uncommon. While prairie voles do form monogamous pair bonds, only living with and raising young with one partner, they aren't totally faithful when it comes to sex. When the opportunity arises, the voles will have sex with a stranger, and it is estimated that about 10 per cent of young are raised by a father they are not biologically related to. And this means there is room for some males who don't settle down, instead living the bachelor life, mating with the occasional female and leaving another male to help her raise his kids.

Larry Young, a researcher at Emory University US, discovered that these 'wanderers', as they are often called, had differences in a gene for vasopressin receptors. This gene has an area where the genetic code repeats a number of times. If an animal has fewer repeats than average, it will end up with fewer vasopressin receptors in certain brain areas, and be more likely to be a wanderer. The more repeats, and the more receptors, the more strongly bonded the animal will be to its partner.

The cuddle chemical

In humans too, it seems likely that evolution hijacked one system, which allowed bonding between mothers and their infants, and tweaked it to create pair bonds between adults. And while this might sound far-fetched, it is actually not unusual. It's much easier to co-opt a system already in use than to build one from scratch. We can see the similarities when we look at the brain, as both the maternal and partner-bonding systems involve dopamine in the 'reward' areas of the brain, and the effects of oxytocin and vasopressin.

In humans, it is harder to pin down the precise effects of these hormones. While we can measure the levels in someone's blood, we can't know for sure that this links to the levels in their brain. And, of course, we can't inject hormones directly into the brain to study their impacts either!

But there are some hints that they play a role. One study, for example, found that men given oxytocin rated pictures of their wife as more attractive, but didn't change their ratings of images of other women. At the same time, their nucleus accumbens was more active when looking at their partner after being given oxytocin, suggesting they found her image more rewarding.

Then there is the fact that, just like prairie voles, humans have receptors for oxytocin and vasopressin in a range of brain areas that have been linked to love, like the reward system and the limbic system, including the amygdala. These two hormones are structurally very similar, and in some cases seem to be able to activate each other's receptors, which makes it hard to untangle them. And they are both important in forming long-term bonds. But looking at human and animal data together, it seems as if they have slightly different roles. Oxytocin is a signal of safety, reducing anxiety and so allowing you to stay in one place and get close to someone. Vasopressin is involved in protecting a relationship, and can lead to aggression and jealousy.

While both men and women make and respond to both hormones, when it comes to bonding men seem to rely more on vasopressin and women more on oxytocin. The way they affect the brain is also influenced by sex hormones like oestrogen and testosterone, which may explain these differences. But there are differences in these bonding networks between individuals of the same sex, too. Partly this is affected by your genes. One study found that men with genes which produce fewer vasopressin receptors are more likely to have marriage problems and report lower levels of satisfaction in their relationships. And differences in dopamine response in the reward system can also have an effect. But this doesn't mean cheating is 'hard-wired'. In fact, as we have seen throughout this book, our brains can be changed by experience, especially when we are young. This seems to be particularly true for our bonding systems.

As we saw in Chapter 2, stress during early childhood can have huge knock-on effects on a person's brain, and some of this may be due to changes in this system. Too little oxytocin during early development because of abuse or a lack of relationships (like the experiences of the children in the Romanian orphanages) can change the way the brain produces vasopressin and oxytocin later in life. It may also affect where in the brain receptors for these chemicals are found, and how many of them are present. This could explain the orphans' issues forming strong relationships later in life, as well as some of the mental health issues they experienced.

The link with mental health is because oxytocin helps you deal with stress. We have long known that people with healthy, supportive relationships (whether romantic or not) are more resilient to negative life events, and they are even less likely to fall ill. It now seems that oxytocin might be part of this, as it can damp down the fight or flight response, which, as we saw in Chapter 4, can cause all sorts of problems if it is activated in the long term.

So does all this mean that some people are doomed to failure when it comes to relationships, because of their genes or their early life experiences? I think that is unlikely. Everything we know about the brain and its complexity suggests that our destiny is not written in our genes, or even in the wiring of our neurons. The chemicals that drench our brains allow for flexibility. And we know that in other areas of our lives we can resist temptation, and overcome our instincts. It may be that there are some people for whom monogamy is the easy option, and others who will always find themselves drawn to the excitement of a new relationship. But acting on that urge is a choice, as is resisting it. And making that choice, over and over again, may well begin to change the brain, making it easier to resist in the future.

Even for those of us who don't see temptation around every corner, being in a long-term relationship isn't easy. In those heady early days of falling in love, our brain chemicals overload our brains with emotions and make us feel like the object of our desire is everything we need. But over time, these feelings fade, as the hormones of long-term bonding take over. These aren't produced automatically, and nurturing them, and your relationship, is an active process. Science has shown that oxytocin is produced when you feel close to your partner, either physically, through sex or cuddling, or emotionally. But if you are in a long-distance relationship, don't worry! Though there is little research, a study by Leslie Seltzer at the University of Wisconsin–Madison found that just talking to someone you love (in this case a child speaking to their mother) can cause oxytocin release.*

If the relationship begins to drift apart, as it so easily can when both partners are living busy lives, oxytocin levels can drop. Maintaining that closeness throughout the years is key to keeping the relationship alive. What all this means is that a relationship is a choice, and one you have to make over and over again. When I fell in love with Jamie, it wasn't a conscious decision, but building a life with him has been. And I hope we will continue to choose each other in the future, encouraging those brain chemicals that keep us bonded. Or, as author Ursula K. Le Guin wrote in *The Lathe of Heaven*: 'Love doesn't just sit there, like a stone, it has to be made, like bread; re-made all the time, made new.'

---

*You do need to pick up the phone, or hop on a Zoom call though, as instant messaging didn't have the same effect.

CHAPTER NINE

A Pain in the Brain

The most pain I have ever experienced was while I was on holiday from university. I was on a ski trip, organised by my college, and one evening we went tobogganing. It was cold, and dark, so we bundled up in as many layers as we could wear, and traipsed to the top of the piste, dragging our small, plastic toboggans behind us. Then we were off! It was a lot of fun at first. The gentle slope was lit by small lamps, and we whizzed down laughing and chatting, using our feet and the (fairly ineffectual) brakes on the sledges to stay in control of the speed. At one point, a German man wrapped in fairy lights skied past, giving us a cheery wave as he disappeared down the mountain ahead of us.

Then we rounded a corner to see what looked, to me at least, like a cliff's edge in front of us. This part, the guides told us, we would take one by one. I was not keen. Unlike some of my friends I was not emboldened by alcohol, and my fight or flight reflex kicked in, full pelt. My heart raced as it got closer to my turn. I considered refusing, but quickly realised there was no other option. I had to get down the slope.

It was probably this fear that caused what happened next. One after another, my friends tumbled off their toboggans part-way down the slope, arriving cold and perhaps a little bruised, but safely at the bottom. Then came my turn. As I started to pick up speed, I planted my feet, and pulled the brakes, trying to regain control. What happened next is a bit of a blur. I must have hit a bump, because suddenly I veered off sideways, hanging on to the sledge as tightly as I could (when really, in hindsight, falling off would have been a much better option. Turns out my survival instincts leave a lot to be desired!). A few seconds later I found myself lying in the snow, on my front, limbs askew, in a position I remember thinking would have made a perfect chalk outline in a murder mystery show. My sledge had hit the wall of ice that divided the slope we were on from the next one, and I had bounced off it, landing a metre or so away.

Interestingly, I don't remember feeling a lot of pain at the time. In fact, I don't even remember it being particularly painful when the emergency services arrived and strapped up my rapidly swelling ankle and wrist. They were a bit sore, which is how I knew these were the two areas that the doctor should X-ray when we arrived at the tiny clinic. But not the kind of pain I would have expected from two broken limbs (which turned out to be the diagnosis). The pain came, however, when the doctor picked up my wrist in both hands and gave it a sharp tug, like he was pulling a Christmas cracker. It was necessary, to realign my bones so they could heal properly, but wow, did it hurt.

I remember crying out, then quickly dissolving into laughter, almost delirious with the pain. But while it was unpleasant, that pain was important. If my ankle hadn't hurt when I put weight on it, I might have walked back to the hotel, unaware of the damage I had done to it. And if my arm hadn't throbbed throughout the following days, I might have been less careful when moving it, and not given it the proper rest it needed to repair. Pain is the body's warning signal. It tells us that we are doing something that might cause us harm, and encourages us to stop. And while it is far from enjoyable, this makes it vital. But there are some people who don't experience pain, and this can lead to all sorts of problems.

The people who feel no pain

Imagine meeting a child who constantly seems to be injured. The first sign something was wrong occurred when they were teething, and managed to almost chew through their tongue. Then, as they got older, they were always covered in bruises, cuts and burns, and regularly in the hospital for broken bones. You might, understandably, think this was the fault of the parents for not taking good care of their child. And, distressingly, it is incredibly common for children who can't feel pain to be taken into care before their disorder is discovered. The inability to feel pain isn't the blessing it might sound – it is incredibly dangerous. As well as injuring themselves without realising it, many people with this condition are unable to feel internal pain, so they are unaware of illnesses like appendicitis, which can, and often do, end up being fatal.

In many cases, the inability to feel pain is caused by a genetic change that affects the development of the nerves that carry pain signals. This means pain signals can't be transmitted properly to the brain. There is, currently, no way to cure this condition, so sufferers have to learn and remember what is dangerous and what is safe, and carry out manual checks on themselves daily to discover any injuries. In some cases, however, it isn't the wiring that's the problem, but the chemicals involved in sending the signal. For these people a treatment might be possible, but to understand how that might work, we are going to have to start with the basics of how pain signals are transmitted.

On the surface, pain seems like a simple phenomenon. In a famous sketch, seventeenth-century philosopher René Descartes illustrated his understanding of pain, showing a man kneeling, with his foot in a fire. Lines connect the foot to his brain, and Descartes suggested that these were nerve fibres, which carried information about skin damage to the brain, causing the sensation of pain. This dogma is still very much alive today.

The problem isn't that it is wrong, as such, but that it is overly simple. Descartes was correct in many ways, and the basic anatomy of pain does follow his image. If you hurt your foot, nerve fibres in the skin send signals to the spinal cord. Here, they release two chemicals into the synapse: glutamate, an excitatory neurotransmitter involved, as we have seen, in just about everything the brain does, and substance P, which is found throughout the body and has a role in promoting inflammation, amongst other things. These chemicals activate other nerve fibres, which carry the signal to the brain.

Here is where it gets a bit more complicated, because if this was the only process involved, cutting those nerve fibres should completely remove someone's pain. But it doesn't.

So what is going on? To answer this, we need to understand the physiology and biochemistry of pain at the most basic level. Pain is a term which can be used to refer to a whole host of different types of experiences and events, so let's start with what seems like the simplest example, and the one Descartes himself used: damage to the skin.

Our skin is far from just being a passive barrier, keeping our organs in and the outside world out. In fact, it contains a host of sensors that can detect everything from vibrations to temperature. It also contains nociceptors, sometimes called 'pain receptors'.* Nociceptors are free nerve endings, which means they don't have any complex structure, they are simply the dendrites of sensory neurons.

Researching this chapter reminded me of a practical class I took part in for my neuroscience course at university. It was held in one of those old-fashioned labs, all wooden benches, covered in burn marks from Bunsen burners and stains made by students generations ago. It's the smell of those labs that will always stay with me – not unpleasant, but distinctive; warm yet potent, something modern wipe-clean labs seem to be lacking. We were put into groups, and told that today's practical would investigate pain perception. I was paired with two other women from my college, Steph and Kat, and after some debate, Kat bravely volunteered to be the subject of our task. Hesitantly, Steph and I blindfolded her and after lifting her arm to drain the blood out of it, fixed a blood pressure cuff around it, tightly enough to cut off her circulation.

What followed was one of the strangest 20 minutes of my time at university. At two-minute intervals, Steph

* Although as you will see later in the chapter, this is perhaps a misnomer.

and I had to stick pins in our friend's hand, and ask her to locate and describe the pain. At first, her descriptions were as you might expect. When we jabbed her index finger, she immediately said 'ow' and described a pricking pain in that finger. Nerve endings in her finger were detecting the damage we were causing, and transmitting warning messages to the brain. These messages were travelling down A-delta fibres, which are large, and coated in insulating myelin, meaning their signals travel rapidly. They detect mechanical damage over a certain threshold (and possibly heat damage as well in some cases) and the pain they produce is described as sharp, stinging or pricking. It is also easy to localise. You know almost immediately exactly where that drawing pin you stepped on has embedded itself in your foot!

But as she sat there, with the cuff blocking the blood flow, her arm began to go numb. The large A-delta fibres, which need a lot of energy, were some of the first to stop working. The next time we pricked her, she didn't respond right away. Then, after a short delay, she told us she could feel pain, but it was different; less of a sharp stinging sensation, instead a duller, burning feeling. And she wasn't sure exactly where on her hand it was coming from. This was because she was no longer receiving signals from her A-delta fibres; instead, the sensation was coming from another type of nerve, known as C-fibres. These are smaller, and unmyelinated, so transmit their pain signals more slowly, and the pain they produce is duller, more of an ache or a burn, and harder to localise. These fibres, being smaller, don't use as much energy, so can survive longer without blood flow.

C-fibres have endings that respond to a range of different potentially dangerous stimuli. Some respond to mechanical injury, like cutting your finger or stubbing your toe. Others are triggered by heat, and still more by chemicals, either those in the environment, or those produced by damaged cells. When our skin is damaged, cells are broken open and begin to leak their contents into the surrounding area, and the damaged cells also release a whole host of chemicals including histamine, bradykinin and prostaglandins. These kick-start the healing process by producing inflammation and causing fluid to leak from blood vessels into the tissue, making it swell. The same process attracts white blood cells to fight infection.

But the chemicals themselves also trigger nerve endings to send more warning signals. This is the longer-lasting pain, often accompanied by swelling and redness, that we experience after hurting ourselves. Some of these chemicals can sensitise nerve endings, making them respond to stimuli that they wouldn't have reacted to before. There are other nerve endings known as silent nociceptors, which don't respond to noxious stimuli initially, but once the area is inflamed, begin responding. This is why a gentle tap to an area around a cut, or the brushing of clothes on sunburned skin, can be so acutely painful. The process is known as peripheral sensitisation.

It is here that some of the over-the-counter painkillers you can buy work their magic. Aspirin and related compounds, for example, block the production of prostaglandins, reducing this sensitisation. This is why these painkillers are also effective at reducing inflammation. However, they only work on this secondary mechanism, and do nothing to block the pain signals caused by the initial injury, so they don't inhibit pain entirely.

Tell me where it hurts

So damage to the skin can cause pain, but it turns out the pain you feel isn't a direct response to the amount of damage. This idea was first introduced in 1962 by Ronald Melzack and Patrick Wall. They put forward the idea that the pain pathway had points along it, or 'gates', that could be influenced by other factors. For example, the pain fibres aren't the only nerves that have synapses in the spinal cord, as other sensory information also travels to this region. These two types of sensory information, they suggested, can interact by means of inhibitory interneurons.

Interneurons are small neurons which go between two other neurons. Inhibitory interneurons, when active, block the signal from being transmitted along whatever neurons they synapse with. When pain signals reach the spinal cord, the interneurons are inactivated, allowing the signal to pass through the 'gate' and carry on to the brain. But if there is a lot of activity in the fibres that carry information like touch, pressure and temperature, this *activates* the interneurons, making it harder for pain information to be sent to the brain, effectively 'closing the pain gate'. So it isn't just the amount of damage detected by the skin that is important, it is also the ratio of these 'pain' signals compared to other sensory signals. And that is why rubbing a bumped knee helps. Providing that extra tactile stimulation can help shut the gate and make the injury feel less painful.

Another interesting finding is that pain to one area of the body reduces responses to pain in other areas. The mechanism for this also occurs in the spinal cord. Neurons here are inhibited when a painful stimulus is applied to any area of the body other than the one they respond to, preventing pain signals being sent from multiple areas at the same time. This explains why when I had my tobogganing accident I didn't notice the huge bruise on my hip until we got back to the hotel and I got ready for bed. My broken ankle and wrist must have been producing enough pain signals to override those from what would in any other situation have been an extremely tender bruise.

We have seen how neurons in our spinal cords can influence pain signals before they even reach our brain, but, of course, the brain has a role to play in changing our experience of pain as well. Think about the mother who lifts a fallen tree off her child, with no thought for her own broken leg. Or the irritating cut that becomes intolerable pain the moment you go to bed and have nothing to distract you from the sensation. These examples can happen because the experience of pain is a brain response, and how our brain responds to messages from our body can be influenced by a whole host of factors.

One way to separate the injury from the experience is by looking at the brain activity of someone in pain. This can also help us start to tease apart what the different brain regions do. While there isn't a specific brain region for pain, there is a combination of areas that have become known as the 'pain matrix'. These areas include the somatosensory cortex, the area of the brain that maps touch to parts of the body. Researchers have found that activity here correlates with where the pain is, and how intense it is. Other important regions are those involved in emotion like the insula, amygdala and hypothalamus. These are what provides your reaction to the pain, and your experiences of its unpleasantness and emotional content.

In brain-scanning studies, regions including the anterior cingulate cortex* seem to be more active when the individual experiences the pain as more unpleasant. If the person is asked to pay attention to the pain, this area becomes more active, and when they are told instead to concentrate on some music, it quietens down. We can also see how expectation can influence pain. If we are told something is going to hurt, we will find it more painful than if we aren't expecting it,† and this area of the brain will be more active.

This is known as 'top-down control', in which our brain can dial up or down our experience of pain based on all sorts of factors. But how exactly does this process work? Scientists are still working on the details, but we are starting to understand the basic process. And it all started thousands of years ago, with a very special flower.

The painkilling poppy

Opium, derived from the humble poppy, has been used to treat pain for thousands of years. There are records of its use as far back as 3400 BC in Mesopotamia and cultivation spread to the ancient Greeks, Egyptians and Persians. In the 1300s, it fell out of favour in Europe, but around 200

---

* This is a region of the brain found deep within its front half, which is involved in empathy, error detection, the regulation of emotions and, importantly for this chapter, the experience of pain, both mental and physical.

† Which means I should probably be grateful to that doctor for the lack of warning he gave me before resetting my wrist!

years later, Portuguese traders discovered the effects could be felt much faster if the substance was smoked. By the seventeenth century, opium had become a source of pleasure and an aphrodisiac. Its use as a medicine also reappeared, with the introduction of laudanum, a mixture of opium and alcohol that was marketed to relieve all sorts of issues, and was even given to babies.

Meanwhile, opium had been introduced to China and East Asia. Realising the drug could be harmful, in 1799 the Chinese Jiaqing Emperor banned opium entirely. But despite this, the opium trade didn't slow down, for one major reason: the British love of a cup of tea. In the eighteenth century, the East India Company imported huge amounts of tea from China. The problem was that China didn't need much from them in return, so the traders had to pay in silver, something they weren't happy about. But China was buying a little opium. So the British ignored the Emperor's ban, continued growing poppies in India and shipped opium to China to trade for tea. Addiction continued to ravage the country and the conflict led to two Opium Wars, 20 years apart. Facing both Britain and France, the Chinese army was defeated both times and forced not only to re-legalise the opium trade, but also to hand over Hong Kong to British rule.

Despite its downsides, opium was, at the same time, revolutionising healthcare. In 1803, the active ingredient was isolated, and named after Morpheus, the Roman god of dreams. Morphine, and other opiate painkillers, are still used today and are the most effective pain relievers known. But they have side effects, such as drowsiness, nausea, constipation, dizziness and slowed breathing, which can be fatal if left unchecked. They also have the potential to be addictive. Throughout the nineteenth century, as well as being used as a medicine, opium and its variants became popular recreational drugs, because they can also produce feelings of euphoria, relaxation, warmth and calmness.

By the end of the nineteenth century, the hunt began for a version that would retain its painkilling benefits without the problematic side effects. German chemist Heinrich Dreser, working for the Bayer Company of Elberfeld, produced a compound called diacetylmorphine,* which was an even more effective painkiller, and which, the company believed, would be less addictive than morphine. Because of its potency, it was marketed under the brand name 'heroin', from the German *heroisch*, meaning heroic or strong.

At first, it was welcomed by the medical community, and sold mainly as a cough suppressant, but we now know that heroin is actually *more* addictive than morphine, because it is able to cross the blood–brain barrier more rapidly. Once in the brain, it is converted to morphine, producing the same effects, but faster. This means it provides more of a 'rush' than morphine does. Soon heroin addiction became its own problem. In the 1920s, regulations limited the sale of opioids, meaning they could only be prescribed by doctors. Despite this, addiction to opioid street drugs like heroin, and to prescription painkillers, remains a huge problem around the world.

* Although it had actually been synthesised for the first time years earlier, it was Bayer who produced the compound commercially.

Pride and prejudice

Although we have known about opium and its effects on the body for thousands of years, there was still a mystery to be solved. How exactly did these compounds reduce pain, and produce their other effects?

Around the beginning of the 1900s, a theory emerged that all drugs must have a corresponding receptor. By binding to these specialised sites on cells, and either activating or blocking them, they could directly change the function of the cells, or influence how our body's natural chemicals affect them. It took until the 1960s, however, when the first drugs were developed to block receptors (beta-blockers, which affect one of the adrenaline receptors), for this theory to be accepted by the mainstream scientific community. Excitement began to grow for the idea that all sorts of drugs might have their own receptors.

In 1972, Candace Pert was a graduate student, working in the lab of Solomon Snyder, at Johns Hopkins University, US. Having spent the previous summer in hospital, taking large amounts of painkillers because of a broken back, Pert jumped at the opportunity to study in more detail the opioids she had been given. She began her research project, attempting to find the sites where they bound to cells – the opioid receptors. To do this, she used morphine labelled with radioactive tracers. Hopefully, this would attach itself to the receptors, so she could see where they were. Unfortunately, as is often the case in science, it didn't work. The signals were too weak, and too noisy, to prove anything. Pert was convinced she was on the right track, but Snyder wanted results, and when her project wasn't delivering, he decided to move her to another, which was more of a 'sure thing'.

But Pert wasn't going to let her research go that easily. She had a hunch that the problem with her results was down to using morphine. She reasoned that she needed something that would bind more strongly with the opioid receptors, so she could see them more clearly. She settled on naloxone, a drug used to rapidly reverse opioid overdoses. Because it worked so effectively, naloxone was assumed to bind to opioid receptors more strongly than morphine, blocking the actions of opioids already in a person's system, and even throwing them off receptors they have already bound to.* But how could she get the drug, without her supervisor's approval?

Luckily, Pert's husband Agu worked in a psychology lab nearby, and was able to get her some naloxone. Once acquired, she sent it to be labelled with a radioactive tracer. She was all set. One Friday, she waited until her colleagues had gone home and she had the lab to herself (with the exception of her five-year-old son, who was happily entertained playing with empty bottles), and started the experiment running. On Monday morning she raced into the lab early, and was overjoyed to find her hunch was right and the experiment had worked. There it was. The signal that showed she had found the opioid receptor.

In her book *The Molecules of Emotion*, Pert describes the world of science as a cut-throat race to be the first to discover, and publish, new findings. Other labs working

---

* Naloxone is sold under the brand name Narcan and carried by emergency services to rapidly treat opioid overdoses.

on opiates were rivals, rather than potential collaborators, and even within her lab she felt the need to fight for her position. Partly, it seems, this was down to the ambitions of her supervisor, and partly to her being a woman. She recounts tales of women scientists being rejected for grants, and of the stereotyping and derision she saw behind the scenes in her male-dominated lab. It was this environment that led to the scandal that was to follow.

By 1978, Pert had been working at the National Institute of Mental Health for three years, leading a team of up to 10. Despite Snyder asking her to promise to stop working on the opioid receptor when she left his lab, she hadn't given up her passion, and was continuing to try to map these receptors to specific regions of the brain. So she was surprised to hear from Snyder, when he called her out of the blue to invite her to an award ceremony. She was even more surprised when she discovered Snyder was receiving the prestigious Lasker Award for his work on the opioid receptor. Alongside him, two other researchers would be recognised: Hans Kosterlitz and John Hughes, who had been the first to isolate molecules from a pig's brain which acted like morphine. Pert was outraged at being left out of the award, and her anger was only heightened by Snyder's suggestion she could 'stand for a bow at the ceremony'.

Determined not to let what she saw as an 'old boys' club' win, she wrote a letter to Lasker, explaining her claim, and her anger and frustration at not being recognised. She also mailed a copy to *Science* magazine.

A few weeks later Julius Axelrod, Snyder's mentor, came to Pert with a request. He was in the process of nominating Snyder, Kosterlitz and Hughes for the Nobel

Prize (it is very common for those who win the Lasker to go on to win a Nobel), and he needed her help. Still furious, Pert refused. She recounts realising at this point why she hadn't been included. The Nobel can only be shared by three living scientists, so one of them had to be left out.

All this controversy, of course, had an impact, and the three men never won their Nobel Prize. But it also had a lasting impact on Pert. She developed a 'notorious reputation', and recounts being 'blackballed', with her invitations to speak at conferences and events taking a dramatic nose-dive. But despite this response from the establishment, the women she worked with, she says, hailed her as a heroine.

She continued working on peptides (small molecules, like opioids) and their receptors, publishing over 250 papers over the course of her career. But she also made an unusual turn into the world of alternative medicine and spirituality. Perhaps it isn't surprising that she moved into a world that was far more welcoming, and less cut-throat. But while the first half of *The Molecules of Emotion* recounts her experiences in science and the findings she made, the second half makes some incredibly non-scientific leaps in reasoning. It claims, for instance, that her discoveries proved that all illnesses had psychological roots (she even mentions how important it was for her to write that letter to Snyder, because not doing so might set her up for 'depression and maybe a cancer or two down the line'), and even made the baffling claim that 'God is a neuropeptide'. It's sad to me that Pert, who was obviously an incredibly passionate and talented researcher, was driven out of the scientific world because of politics and prejudice. Who knows how many other brilliant scientists we have lost in the same way, but who went quietly, rather than making their story heard as Pert did.

Don't worry, it's just a flesh wound!

Thanks to Pert's tenacity, we had evidence for opioid receptors in the brain. We now know that there are receptors for opioids at various points along the pain pathway. Starting at the source of the pain, there are receptors on the sensory neurons that carry that pain information to the spinal cord. When opiates bind to these neurons, they make it harder for them to be activated, meaning they release less of their neurotransmitters, glutamate and substance P into the synapse at the spinal cord. Without chemicals to carry the signal across, the second neuron, in the spinal cord, never receives the message, and no pain can be felt. But opioids don't stop here. There are also receptors on this second neuron, so even if the message passes across the synapse, opioids here make it less likely the second neuron will fire, again preventing the signal from reaching the brain.

But just in case the signal does make it up the second neuron to the brain, opioids have another role to play. As well as the ascending pathway, which takes pain information to the brain, there is another pathway, going in the opposite direction, which interacts with it. Opioids act on a region of the brain called the periaqueductal gray (PAG), to activate this descending pathway. Once activated, the descending pathway releases chemicals including GABA, serotonin, noradrenaline and more opioids into the synapse between the sensory neurons and the neuron that would carry the signal up the spinal cord. These chemicals act together to make it harder for the signal to be transmitted, so the feeling of pain is reduced.

Imagine your pain signals as tiny ambulances, travelling along the road which is your ascending pain pathway, and your descending pain control pathway as a river, held back by a sluice gate. When released, the river will flow down the hill and flood the road, stopping any traffic from getting through. Close the sluice gate again, and the water will dissipate, allowing the ambulances to resume their journey. So opioids, by releasing the river of descending pain control, can have huge knock-on effects.*

This is what makes morphine and related compounds such powerful and effective painkillers. But the reason they work is because our body produces similar molecules itself, so-called endogenous opioids like enkephalins and endorphins. And it can use them to block pain. Indeed, some of the areas of the pain matrix, that network of brain areas we discussed earlier, might actually be involved in the control of pain, rather than the experience of it. Activation in the anterior cingulate cortex, for example, seems to link to increased activity in the PAG, and electrically stimulating the PAG can block pain signals. This suggests the cortex might be able to dial up or down pain using this system.

For example, we know that people who *worry* more about pain (known as pain catastrophisers) actually *feel*

* If you wanted to take the analogy one step further, you could say that as well as opening the sluice gate, opioids also slash the tyres of the ambulances (as they block signals from the sensory neurons) and turn the road surface to treacle (by making it harder for the spinal cord neurons to fire). But that might just be analogy overload!

more pain. Studies have found that people who have negative beliefs about pain, such as thinking they can't cope, or that it will never get better, rate the same stimulus as more painful. The emotional areas of their brains are also more active. We don't know exactly why this is, but there are a few ideas. It may be that these thoughts mean they pay more attention to the pain, and can't distract themselves from it so easily, so it feels worse; or it may be that they expect it to be more painful and, as we have seen before, expectations can influence experiences. It seems that regions of the cortex, including prefrontal areas and the anterior cingulate, are able to exert 'top-down control' over the pain system, using our natural opioid molecules and this descending pain control pathway. Changes in attention or expectation may alter this process, making pain more or less painful.

In fact, in some cases it is this system that is dysfunctional in those people who can't feel pain. They have overactive descending pain control pathways, so their 'rivers' are always flowing, and their 'ambulances' can never reach the brain. This was proved by giving naloxone to one of these very rare individuals, blocking their opioid receptors. Within minutes they were able to detect painful stimuli that previously they couldn't feel at all, and even reported aches in a leg that they had broken several times in their life. There is hope that by studying these people, we may be able to develop better drugs to treat pain, but in the meantime, are there any ways we can activate our own pain-blocking pathway?

It turns out there are. One common experience that shows the power of the body's natural opioids is something known as 'stress-induced analgesia'. In certain cases, when we are under intense stress, the brain activates the descending pathway, blocking pain and allowing amazing feats like soldiers walking miles on a broken leg to get to safety.

This process starts with neurons in the cortex, amygdala and hypothalamus. Activation of the stress system in these regions causes the release of endogenous opioids and cannabinoids (natural chemicals related to the active ingredient in cannabis) in the PAG. Just as we saw before with morphine and related drugs, this activates the descending pathway, blocking ascending pain signals.

But, as ever, we must be cautious interpreting these studies. The extent of stress-induced analgesia varies depending on the type of pain, the type of stress and between individuals. We know, for example, that distraction can also act as a painkiller. People with severe burns who are given VR (virtual reality) headsets to wear while having their dressings changed reported lower pain levels than without the headsets. And people can learn to distract themselves from pain. In an imaging study, led by Irene Tracey, head of the Department of Clinical Neurosciences at the University of Oxford, people managed to reduce their pain through distraction, rating the same stimulus as less painful than when they were paying attention to it. When they distracted themselves, their PAG became more active, supporting the idea that distraction works through the same mechanisms as stress-induced analgesia. Ascending pain neurons in the spinal cord are also less active when someone's attention is diverted elsewhere than when that person pays attention to the same pain stimulus.

And naloxone, which blocks the opioid receptors, weakens the pain-reducing effects of distraction, showing that the distraction too relies on natural opioids.* So how can we know that stress isn't simply acting as another kind of distraction?

D*mn that hurts!

One scientist who has been trying to tease apart this problem is Richard Stephens, Senior Lecturer at Keele University, UK, who works on a rather surprising topic: swearing. Stephens's research was inspired by his own experiences, and the fact that everyone he knew tended to swear as a reaction to getting hurt. Ever curious, he wanted to know why. Was it just a learnt response, could there be a benefit to it, or could it even make our pain worse? I asked him to explain how something as simple as swearing could affect our pain levels:

> Swearing may have been a form of catastrophic thinking… Catastrophising is the understanding of how much threat an injury carries. We are all on a spectrum, but someone who thinks a paper cut might be life-threatening is catastrophising, and it makes the pain worse. So, if swearing was a form of catastrophic thinking, it would actually be making the pain worse. But this struck me as unlikely – when I swear because I'm in pain, I'm hoping the pain is going to get better.

* Although there was still a slight pain-reducing effect of distraction when patients took naloxone, so there must be multiple mechanisms at play, some of which don't rely on opioids.

He set about devising an experiment. One of the most common ways researchers measure someone's pain tolerance is asking them to keep their hand in a bucket of iced water for as long as possible. As you can imagine, this starts off uncomfortable, but rapidly becomes incredibly painful. The good news, though, is that as long as a safe maximum time limit is stuck to, there is no long-term damage, so it's easier to get through ethics boards than other methods of hurting volunteers would be!*

Over the next few years, Stephens spent hours with his participants, recording how long they could keep their hands under when swearing, and comparing this to when they were saying a less rude word.

> We get people to nominate a swear word they might use, and then as a control condition, we still want them to be articulating in the same way, so we ask them to give us a word to describe a table. Then they repeat the swear word or the table word once every few seconds.

They found over a number of studies that swearing does allow participants to keep their hand in the iced water for longer than the neutral word. It also works better than words Stephens's team have made up, which sound a bit like swear words: words like 'twizzpipe' or 'fouch'. But they didn't have the same painkilling effect as the real swear words.

In some of their studies, Stephens has found swearing seems to cause an increase in heart rate, and swear words are rated as highly emotional. Stephens thinks that it is

* Paper cuts are off the table, apparently…

this emotional response that causes the pain relief, as swearing is triggering stress-induced analgesia. Interestingly though, it doesn't work for everyone: 'How often you swear in everyday life has an effect… the people who swear every day don't get as much benefit.'

Amazingly, swearing can even help reduce mental pain. Researchers using a game called Cyberball, where subjects rapidly become excluded from a virtual game of catch, found that the feelings of ostracism were reduced when subjects swore. Again, those who swore more in their normal lives showed less of a benefit, suggesting a similar mechanism might be at play here. So, should we all take to turning the air blue when we stub our toes? Well, as Stephens puts it, there are few downsides:* 'One of the things with swearing is, it's free, it's drug free, it doesn't have any calories, and it does carry these benefits.'

Painkilling placebos

We have seen that a whole range of different influences can affect how much something hurts, from your environment to your personality. But, to me at least, one of the most fascinating things about pain is the placebo effect. We are all open to suggestion, however much we might want to deny it, and studies show that this is particularly true when it comes to pain. While the first thing that comes to mind for most people when you say the word placebo is a sugar pill,† placebos can actually come in a range of forms.

---

* Assuming, of course, you aren't a primary school teacher, or on national radio at the time.

† Or the excellent 90s rock band.

A placebo is defined as an inert substance or treatment, designed to have no therapeutic value. But this definition is inherently flawed, because we know placebos *do* have a therapeutic value. That's why every new drug must be tested against a placebo, so we know that any improvements are down to the drug itself, not simply the act of taking a pill or receiving an injection.

So how can an inert substance like sugar or saline reduce someone's pain? It all seems to be down to those opioids produced by our brains. When we are given a pill, or an injection, and told that it will reduce our pain, we have an expectation that that is what will happen. And it seems this expectation is enough to encourage our brains to release opioids from the PAG and into our descending control system, blocking pain signals and providing relief. Studies have shown that blocking opioid receptors can prevent placebo painkillers from working.

If we scan the brain of someone being given a placebo, we can start to pick apart the brain regions involved. Many of the regions whose activity correlates with the strength of the placebo effect are, unsurprisingly, in the parts of the brain that respond to pain, such as the insula and somatosensory cortices. But during the *anticipation* of pain, prefrontal regions are also important. More activity here seems to correlate with less activity in the pain response areas. This could be where expectations are having their effect, as we know the prefrontal areas are important for conscious thought. And, they are connected to the PAG and its descending pain control pathway as well as the striatum's reward system, which has also been implicated in the placebo response.

But, as with everything in the brain, placebos are more complicated than they might initially seem. In most studies, expectation is induced by a scientist lying to a participant. For example, we might see someone's pain rating go down when they are told they are being given morphine, even when it hasn't yet been administered. And their pain might go up again when they are told it has been stopped, even though they are still receiving it (this is actually the nocebo effect – the placebo effect's evil twin). In fact, telling someone you are giving them a drug, but giving them nothing, can be more effective than giving it to them without telling them!

But verbal instruction can't account for everything. For example, the placebo effect works much better in people who have previously been given morphine, and experienced its analgesic effects. This suggests that part of the effect that placebos have is down to a non-conscious process of learning, often called conditioning. Over time, people learn to associate certain experiences with pain relief, and the placebo effect comes into action.

For example, in one study by Luana Colloca and colleagues, participants were told that they would be given the same electric shock each time a light came on. But, the experimenters explained, when they saw a red light they would also experience a procedure that would make pain more intense, while green meant a procedure that would reduce pain. Amber would mean the shock alone. The experimenters didn't actually use a second 'procedure', instead just turning the shock up when the light was red, and down when it was green (amber, with pain between the two, acted as the control). This experience, and the lie, gave the participants the sensation of the lights affecting their pain levels, setting them up for the next part of the experiment.

Once this training phase was over, the participants were tested. This time the *same* pain stimulus was given every time, but the participants rated it as more intense when seeing red than amber, and less intense when seeing green. They also found that the more training the subjects had, the more pain relief was associated with the green light, and the longer the effect lasted.

In these experiments, there is still an element of expectation. But Ted Kaptchuk at Harvard Medical School noticed something interesting about the placebo studies he had been running. He was working with people who had irritable bowel syndrome, many of whom had tried multiple treatments for their symptoms, with no improvement. They often told him they didn't expect the treatment to work, and yet many of them did get better during his studies. So he started to wonder if expectation mattered at all. Might placebos work even when you know they are placebos? His team set about finding out. It turned out they did, and even when patients were told that the sugar pills they were being given had no active ingredients, they often felt better. Scientists don't know exactly how this works, but it seems as if conscious expectations aren't solely responsible for the placebo effect, as we once thought. Instead it might be a combination of factors that work together to produce the effect.

Not only is the fact that placebos can work when patients know they are placebos a fascinating finding for science, it is also of huge clinical relevance. While we know that placebos could help many people, without the negative side effects often produced by painkillers and other drugs, there is a problem for doctors prescribing them, because doctors don't like to lie to their patients. Telling someone they are being given morphine, while actually giving them a placebo such as saline, would be an ethical nightmare, and go against the Hippocratic oath. But if you could tell someone they are getting a placebo, but that many people have still found it helpful, maybe then doctors could use the power of placebos without losing the trust of their patients.

The vicious cycle of chronic pain

One group for whom placebos might provide much-needed relief are people with chronic pain. As we have seen, pain evolved as a warning system, and for most people, this system works well. Pain prevents further damage. It instructs us to remove our hand from the hot pan, or to avoid putting weight on our broken ankle to allow it to heal. It also teaches us not to repeat the actions that hurt us in the first place. Someone who has scalded themselves by spilling a hot cup of coffee on their lap while driving is likely to be more careful next time. Their painful experience has acted as a learning tool.

But in some cases, our pain system can learn too strongly, becoming overly sensitive, and this can trigger chronic pain. Unfortunately, chronic pain is incredibly common. One review study published in *BMJ Open* in 2016 found that between a third and half of the UK population suffers from some form of chronic pain.

The first step towards the development of chronic pain is the peripheral sensitisation we discussed earlier in the chapter. Damage to the cells causes the release of chemicals, which increase the activity of the receptors in the pain system, as happens when there is inflammation surrounding a wound. But if this happens over a prolonged period, longer-term changes can occur in the neurons themselves, making them more sensitive. This is known as central sensitisation.

Another common reason for central sensitisation is nerve damage. When nerves are damaged, they may repeatedly fire. This activates nerves in the spinal cord, which carry the pain signal to the brain. But as well as the nerves that respond *specifically* to one type of input, whether that is pain or touch or temperature, there are also nerves in the spinal cord which respond *non-specifically* to all these sensations. These tell your brain about the strength of the stimulus, firing more rapidly the more intense the sensation is. When they are repeatedly activated, as in cases of nerve damage, these cells begin to change. They become more sensitive, responding strongly to stimuli that before would have produced only a weak response.

This process is very similar to long-term potentiation, a mechanism involved in learning (see Chapter 2). Repeated input from the nerves causes the release of large amounts of glutamate, which, along with other chemicals like substance P, means the second cell is activated for a longer period of time. This triggers changes in the neurons, strengthening their connection and making it easier for the first neuron to activate the second. This can lead to sensations that previously would only have been felt as touch, activating the pain nerves, and so hurting. The nerves themselves may even be able to cross-excite each other, so one signal can soon spread to cause a large amount of pain. There can also be reductions in chemicals like GABA, which normally inhibit these pain signals.

Although this mechanism is best understood in pain caused by inflammation and nerve damage, it seems likely other forms of chronic pain have similar mechanisms. In fact, we know that one risk factor for developing chronic pain after an injury is inadequate pain management. Living for a long time with a painful injury and not enough painkilling drugs puts you at risk of this central sensitisation process. You can think of it as an error in the body's warning system. Repeated activation would normally mean you are continuing to do something that could cause you damage. The body, effectively, thinks you are ignoring its warning systems, so it ramps them up.

The brain also plays a role in chronic pain, as changes occur in the descending pathway. Cells in the brainstem, for example, can trigger the release of serotonin in the spinal cord, which seems to facilitate pain signals, and also noradrenaline, which inhibits them. In chronic pain, these neurons too undergo changes, releasing more serotonin and less noradrenaline, making the nerves in the spinal cord easier to activate.*

* Ten points if you noticed something confusing here. Earlier, I told you that serotonin helped the descending pathway block pain, and now I am saying that it helps activate the ascending pain pathway. This wasn't a mistake. Serotonin seems to be able to play either role, depending on what type of pain is involved. It can do this because it has lots of different receptors – which ones are activated in each case can entirely change the effect it has on pain transmission.

There are also changes in the prefrontal regions of the brains of people with chronic pain. Studies have found a decrease in grey matter in parts of this area, which connect with the limbic system. This may mean these patients are less able to use top-down control to inhibit their pain. One study that followed people for a year after a back injury found that half of them went on to develop chronic pain, and their brains changed throughout the study. So it is clear that the pain causes these changes, not the other way around.

Another way to think about the development of chronic pain links with prediction error and reward learning, which we discussed in Chapter 3. Our brain is constantly making predictions about the world, then comparing them to the outcome, and using any differences (or prediction errors) to adjust its expectations for next time. Say it predicts that bending over will be fine, but actually, it hurts your back. The next time you bend down, you will expect some pain. So far so good. The problem arises because, as we have learnt, expecting pain can actually make it worse. When you bend over again, because of your expectations, the pain *is* worse. That is another prediction error, and your brain ramps up its expectations of pain. You are trapped in a vicious cycle. This seems to happen in the same circuits involved in reward processing and motivation (see Chapter 3), with the cues that predict pain activating the nucleus accumbens. How exactly this connects to the areas of the brain known to modulate pain, however, isn't clear.

So how can we break this cycle? The answer seems to be early, and effective, pain relief. Studies have found that those who had their pain managed early expect that their pain will decrease in the future. But each ineffective intervention increases a person's expectations of future pain.

None of this implies that chronic pain isn't real. Chronic pain is as real and as troubling as any other type, but it is important to understand the mechanism, because for many people, it may no longer be connected to the problem that originally triggered it. While you may feel the pain in your back, or the arm you broke years ago, it may actually be a disorder of the nervous system. This is what needs to be treated, not the body part itself.

This insight could lead to some interesting treatment ideas for chronic pain. Perhaps, if we can't alter the pain signals themselves, we could change how a person's brain responds to them, tapping into that difference between intensity and unpleasantness. Studies have found, for example, that hypnosis reduces activity in the anterior cingulate cortex. Patients report that while they are still in pain, it doesn't bother them as much, and they are more able to continue with their daily tasks.

It is clear that drugs have an important role in treating pain, particularly acute pain. But they do, as ever, have side effects. Findings are mixed, but there is evidence that even the most effective painkillers, the opioids, may stop working if you take them for a long time. This could be because people develop tolerance to them, with their bodies downregulating receptors, or their own opioid chemicals, to counteract the drugs. Understanding more about these processes and how our bodies regulate themselves when flooded with external chemicals could be key to developing better pain treatments in the future.

For longer-term pain, there may also be a role for non-drug treatments. We have seen throughout this chapter how the brain can control incoming pain signals, activating the descending pathway to reduce our pain. Using this consciously is nothing new. If you have ever rubbed a bumped knee, sworn loudly after stubbing your toe or distracted a crying child with a favourite toy, you have used these pain control systems. And pain management, using techniques like cognitive behavioural therapy, yoga and meditation, are already making their way into mainstream medicine. These aim to affect the brain's perception of pain, or treat the mood issues that so often go alongside pain conditions, and can make them worse. But by studying the brain's pain control pathways more, it may be possible to find ways to activate them more efficiently and use them to control pain in more extreme environments.

Of course, as always, there is a balance to be struck here. We have seen that an overactive descending pathway can cause as many problems as an underactive one. Our brains and nervous systems are always trying to stay in equilibrium, not overloaded with too much of each chemical or struggling to communicate with too little. Finding ways to keep them in this state could help us in so many areas of our lives. But I'd also recommend not crashing your sledge into a wall of ice any time soon.

Conclusion

Throughout this book we have seen that the brain is more complex and interconnected than even I had previously realised. So often, processes aren't controlled by a single brain area, or even a single network of areas, but by competing networks which interact and inhibit each other. The balance between these networks, controlled by the chemicals that allow them to communicate, determines whether we are awake or asleep, hungry or full, and decides whether we pursue a goal. And this balance is a delicate one, easily nudged one way or the other by the smallest of changes in the environment, or our behaviour. It is this that gives us our amazing flexibility and allows us to adapt to the world around us.

As scientists start to pick apart these networks, they are beginning to discover just how many different processes our brain chemicals control. By activating different receptors, the same chemical can have seemingly opposing effects, even within one brain region. The rate at which it is released, and how long for, can also change its impact on our brains. This is why the simple idea that one brain chemical is responsible for one type of feeling or behaviour, often peddled by those trying to sell us something, is so far from the truth. Giving up your phone for 24 hours on a 'dopamine fast', which is claimed to help you reset compulsive behaviours like constantly checking your notifications, won't actually lower dopamine levels. And nor would we want it to – low levels of dopamine, remember, cause the symptoms of Parkinson's disease.

This points to another important factor that is often ignored when people are talking about neurochemicals. While too little of a chemical might be bad news, too much can be just as damaging. Low serotonin levels might, in some cases, be linked with depression, but high levels are associated with anxiety. Much like Goldilocks, it seems our brains have a 'sweet spot' when it comes to chemicals, and deviating from this, in either direction, can cause problems.

There might also be different sweet spots in different parts of the brain. Currently, most drugs target the brain as a whole, having their effect wherever there are receptors for the chemical they affect. This means they can be a bit of a blunt tool, but that doesn't mean we shouldn't take them. Currently, the best treatments we have for most mental health conditions include medications, and while we might not understand exactly how every one of them works, they can be very effective. One hope is that in the future we may be able to develop drugs that precisely target the network that is malfunctioning, which should reduce the risk of side effects.

This might sound like a pipe dream, but in another area of healthcare it has already become possible. Traditionally chemotherapy for cancer involves giving the patient drugs that damage cancer cells, but because these are usually delivered via the bloodstream, they travel throughout the body and damage healthy cells as well. This is why chemotherapy can cause so many side effects, from hair loss to nausea and fatigue. Other treatments for cancer, such as radiotherapy, target the tumour precisely, only affecting that region of the body. Now, newer chemotherapy drugs have been developed which target tumours directly. Even when given in the blood, these bypass healthy cells and latch on to the faulty ones, hopefully preventing the cancer cells from functioning.

The exact same techniques as used in targeted cancer treatments are unlikely to work in the brain, but the idea is similar: to devise a way to deliver a drug directly to the malfunctioning area, without affecting healthy cells. But to do this, we have to know which networks to target, in each individual person. This requires not just a better understanding of the fundamentals of brain science, but also a personalised approach to medicine.* Each of our brains is unique, built originally from our genetic blueprint, but shaped by the experiences we have each had throughout our lives. In the future, it may be possible to scan the brain of someone with depression, for example, and determine exactly where their treatment should be targeted. As we have seen, there may be a number of different mechanisms that can lead to depression, so this personalised approach could help diagnose exactly why someone is suffering, and select the most appropriate treatment for their underlying cause.

* While understanding the average brain better may on its own be enough to improve treatments to some extent, I believe the differences between people's brains mean that we will rapidly reach a ceiling in how effective treatments are, until we start individualising them more.

Sadly, this level of personalisation is still a way away, as is developing a method for delivering a drug to just a certain region of the brain. But other treatments are in use which do target particular brain areas. Tiny electrodes, for example, can be implanted in the brains of patients with Parkinson's disease. These deliver electrical impulses, like a pacemaker for the brain, and are often effective in reducing the motor symptoms associated with the illness. These implants have also been trialled for chronic pain, OCD and severe depression, and although only small numbers of individuals have been treated so far, results are promising. However, this treatment requires brain surgery, which is inherently dangerous, so it is currently a last resort, reserved for patients who haven't responded to drugs. Exciting alternatives, using electrical or magnetic pulses delivered through the scalp, are now starting to be investigated. These can be targeted to different parts of the brain, just like the implants, but are non-invasive, making them cheaper and easier to use, and much less risky.

These techniques are exciting because they manipulate our internal networks, activating neurons and changing the ways our natural chemicals are released. But there are other ways of doing this too, by changing our behaviour. We have seen throughout this book that by changing what we do, we can change the activation of brain regions and the release of chemicals. This could be by rubbing a bumped knee to reduce pain, or practising a language in order to store it efficiently in the brain. Our brains are amazingly adaptable, thanks to the chemicals that bathe them, and this is why behavioural techniques can be so effective, from helping us manage stress to getting a better night's sleep. And for more serious conditions, where drugs or surgery might be needed, these behavioural techniques can add to their effectiveness, working together to restore the brain's balance.

The more we learn about the complexities of our amazing brains, the more we can use this knowledge to our advantage. If we know we tend to be tempted by the sight of the biscuit tin, we can hide it in a high cupboard, using our logical, forward-thinking prefrontal regions to override the desires and instincts initiated by the impulsive reward networks. As we learn about the unconscious biases we all fall prey to, we can build techniques into recruitment to ensure we select the best candidate for the job, which may not be the one our biases tell us to go for. And if we know that we have a tendency to become stressed and overwhelmed by the news, we can consciously make the decision to only check it once a day, rather than being pulled into it through phone notifications. Our brain chemicals may cause us to have these underlying tendencies, but they also give us the chance to change and adapt them to improve the way we live our lives.

From moods to pain, we've seen how our brain chemicals affect every aspect of our lives, and how dysregulation of these chemicals can cause all sorts of problems. But we have also seen how they allow us to change, grow and adapt, shaping our incredible, complex brains through our actions, and allowing us to build the brain, and the life, we want for the future. And I think that's worth celebrating.

Glossary

autoreceptor: A receptor for a chemical which is found on the same neuron that releases the chemical. They can be found on the cell body, the dendrites, the axon or the first side of the synapse, where they are known as pre-synaptic.

axon: The long threadlike part of a nerve cell, which carries electrical signals.

blood–brain barrier: A semipermeable membrane which separates the blood from the fluid in the brain. This allows certain molecules, like water, oxygen and some hormones to pass through, but prevents many other chemicals and most pathogens from entering. It also contains selective transporters for necessary molecules like glucose. This allows tight control over the chemicals that make it into the fluid surrounding the brain.

brainstem: The area at the base of the brain which connects it to the spinal cord. It is responsible for vital functions like swallowing and breathing.

consolidation: Processes which stabilise a memory after its intital formation in the brain.

cortex: The surface layer of the vertebrate brain, responsible for higher brain functions. In humans, it is highly folded.

cytokines: Small messenger molecules secreted by a wide range of cells, which affect other cells. Particularly important in the immune system, some cytokines promote inflammation, which can help fight infection- these are known as pro-inflammatory cytokines. Others, called anti-inflammatory cytokines, reduce inflammation.

dendrites: The branching structures at the end of the neuron which receive input from other cells and transmit the signal to the cell body.

endogenous: Something that comes from the organism- endogenous chemicals are those produced by the body.

excitatory: An excitatory neurotransmitter, when it binds to receptors, makes the second neuron more likely to fire.

glial cells: Any cells in the brain that aren't neurons, including those that produce the white matter (myelin).

growth factor: A substance which encourages the growth of cells.

habituation: The decrease in response to a stimulus when it is repeated frequently.

hormone: A molecule that travels through the blood from its release site to a target organ where it binds to specific receptors to change how a cell or tissue functions.

inhibitory: When inhibitory neurotransmitters bind to a neuron, they change the flow of ions, making it harder to activate that neuron.

interneuron: A neuron positioned between two other neurons. Often involved in reflexes.

ion: An atom or molecule which has lost or gained one or more electrons, giving it an electric charge.

limbic system: A complex network of brain areas (or possibly multiple networks) which are involved in emotions.

LTP (long-term potentiation): Long lasting changes in the synapse which make it easier to transmit a signal. Thought to be the cellular mechanism of learning and memory.

monoamines: A class of neurotransmitters grouped because of similarities in their chemical structure. These include dopamine, noradrenaline, adrenaline and serotonin, amongst others.

motor neuron: A nerve cell which sends signals from the brain or spinal cord to a muscle or gland.

myelin: A fatty substance which winds around axons and increases the speed at which messages are sent along them. Sometimes called 'white matter' because of its pale colour.

myelination: The process by which glial cells wrap the axons of neurons in myelin.

neurogenesis: The growth and development of neurons.

neurotransmitter: A chemical made in and released by a pre-synaptic neuron, which affects the functioning of the post-synaptic neuron or cell.

post-synaptic: The second neuron in a pair, separated by a synapse. This neuron receives the signal from the first.

prefrontal cortex: The front part of the brain, situated behind your forehead. This area is important for behaviours like planning, decision-making, and self-control and is particularly large in humans.

pre-synaptic: The neuron which, when activated, releases chemicals into a synapse. These travel across the gap to the post-synaptic neuron.

receptor: 1. A structure in or on a cell which, when a specific substance, such as a neurotransmitter or hormone binds to it, changes in some way. This can convey messages to the cell. E.g. when glutamate binds to glutamate receptors, it triggers changes which make the neuron more excitable. 2. The end of a sensory nerve which responds to a change in stimuli, e.g. temperature.

reconsolidation: The process of storing a memory again after it has been recalled. In some cases, the memory can be altered during this process.

reuptake: The reabsortion of a neurotransmitter by the neuron that released it (the pre-synaptic neuron).

sensitisation: The process of becoming more sensitive to a stimulus.

sensory neuron: A nerve cell which sends signals about external stimuli.

stimulus: Something that causes a change in an organism's environment, which the organism can respond to.

transporter: Transporters are found in the membrane of cells and move chemicals between the inside and outside of that cell.

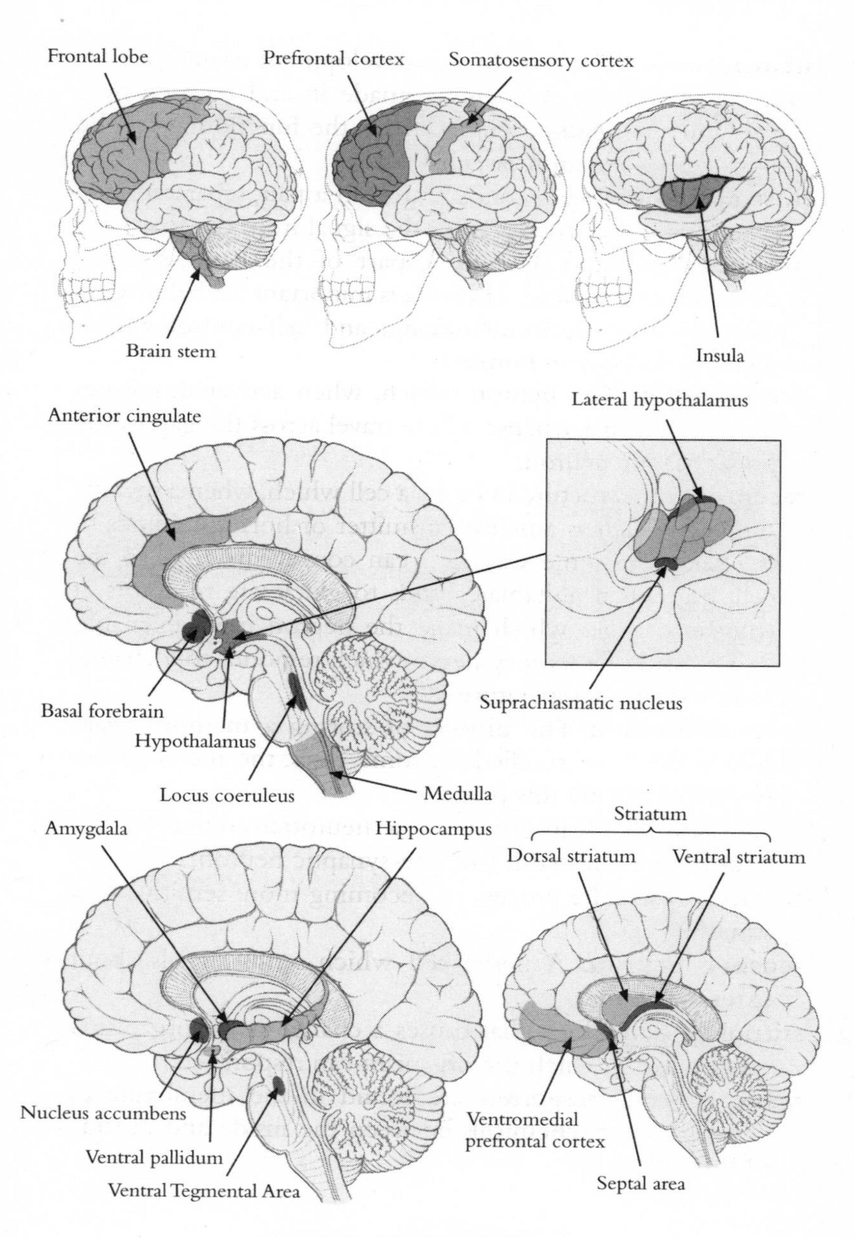

Diagram of the brain

Acknowledgements

Writing this book has been a long process, and it hasn't been a solo endeavour, so I would like to thank everyone who has made it possible:

My husband Jamie – thank you for the lunch-time discussions, the evening pep-talks and the weekends spent reading drafts and discussing changes. And thank you for your willingness to let me talk about our relationship, and our differences – and make the odd joke at your expense! Without your help and encouragement this book simply wouldn't have been possible.

My parents, for the proof-reading, the googling, the advice and the never-ending support and encouragement. And for sparking a love of science and a curiosity about the world that has made me the person I am today.

My friends who so kindly gave up their precious time to read the book at various stages of its development. Emma, Mitch, Trent – each of you provided a different viewpoint, and helped me change the book for the better, so thank you for that. And other friends, especially Kat, who have provided pep-talks, acted as sounding boards and boosted me along the way – thank you for your unwavering confidence in me.

I am forever grateful to those researchers who were happy to speak to me for the book, to review chapters and to answer my incessant emails. Your patience and generosity has been vital in allowing me to write this book. And to all the scientists whose work I have covered in this book – without your dedication, tenacity and passion, we wouldn't understand the brain in the way we do today. Thank you for giving me the chance to tell your stories

My editors Anna and Angelique for your hard work, and to Jim, for believing in me, and giving me the opportunity to write this book.

All the people in my life who have told me during green room chats at various events that I can and should write a book – I remember each one of those and am grateful for them.

And to the drunk woman at a science show who told me to stop doubting myself, and to call myself an author, your words will forever ring in my ears, whenever I find myself hesitating to promote something I have done, or talking it down: "Would a man say that?!". Thank you – you probably don't remember our meeting, but the (slightly aggressive) kindness of one stranger really can make a difference.

Index